AF577187

NANOTECHNOLOGY
USES AND APPLICATIONS

Edited by
Anil Varma

2012

Icfai Books
The Icfai University Press

NANOTECHNOLOGY – USES AND APPLICATIONS

Editor: Anil Varma

First Edition: 2012
Printed in India

Published by

This book is published by IUP.
University Campus, Agartala-Simna Road,
P.O. Kamalghat Sadar, Agartala – 799210, Tripura (West)
E-mail: info@iupindia.org
Website: www.books.iupindia.org

ISBN: 978-81-314-2731-6

Contents

Section III

Progress and Future Outlook

Overview

Nanotechnology is rapidly becoming a ubiquitous technology with a potential to impact every aspect of modern human civilization. Almost every aspect of human endeavour will be affected, such as agriculture and food, communication, computers, environmental monitoring, textiles, robotics, healthcare and medical technology. Indeed, it is in the final two areas that many observers consider that nanotechnology will have its most immediate and dramatic impact.

There is a little of nanotechnology for everyone. However, even with all the media attention, most of us still do not know what nanotechnology really is and why it is important to us. Most people I have talked to only a slight clue as to what it is. I would even venture to say that researchers, even the ones involved in nanotechnology research, do not quite understand it thoroughly even though they probably understand it more than the average person does. There are many articles on how to invest in nanotechnology but not too many explain just exactly what it is to people who need to know.

In the first article "**Nanotechnology for Agriculture and Food Systems – A View**", the authors *H C Warad* and *J Dutta* explain that nanotechnology has the potential to revolutionize the scientific world by allowing scientists to manipulate matter at the atomic or molecular scale using physics, engineering, chemistry and biology. Nanotechnology could be used to revolutionize food and agriculture systems. Novel agricultural and food security systems, cellular biology, environmental protections, disease treatment, delivery methods etc., are just a few examples where nanotechnology could have a signeficant impact. The article highlights that in agriculture and food systems, the fundamental life processes are explored through research in molecular and cellular biology. New tools for molecular and cellular biology are needed that are specifically designed for separation, identification and quantification of individual molecules. This is possible with nanotechnology. It further explains that nanotechnology applications in food and agriculture are in the nascent stage and over the next decade we will see increasing uses of tools and techniques developed by nanotechnology to detect carcinogenic pathogens and biosensors for improved and contamination-free food and agricultural products.

Nanotechnology is receiving a lot of attention from companies, universities and governments. But what does it mean for existing businesses and new businesses in the electronics market. Is it a real tool for today or are the applications, way out in the future? Will it be economic or outrageously costly? In the article "**Real Life Applications of Nanotechnology in Electronics**" the author *Alan Rae* discusses nanotechnology's impact on semiconductors, passive components, display materials, packaging and interconnection. Nanotechnology is like a toolkit for the electronics industry. It gives us tools that allow us to make nanomaterials with special properties modified by ultra-fine particle size, crystalline structure or surfaces. The article highlights that over the coming years we will see significant introduction of nanomaterials and novel production processes based on nanotechnology which will address key issues of importance to the electronics industry. Longer-term use of the technology will allow to meet customer requirements by extending existing technologies or replacing them with new ones.

"**Samsung's Washing Machines: The 'Nano' Innovation,**" the article authored by *Roopa Umashankar* highlights that the application of revolutionary 'nano' technology in Samsung's home appliances paved the way for a new genre of washing machines. Samsung's new washing machines utilized nano-sized silver particles as disinfectant to keep garments germ-free even up to a month after being laundered. With the introduction of the Silver Nano washing machines, the global washing machine market has undergone a tremendous change since the inception of washing machines in the 19th century. The article further explains that the Silver Nano washing machine utilizes a small device containing silver plates near the washtub. When electricity is passed through these silver plates, they emit silver particles, which are then sprayed into the tub during the wash and rinse cycle. This machine coupled time saving along with doing away with the need of hot water to destroy germs. The new technology was also extended so that Silver Nanotechnology would be a major contributor to the expected 50% sales in the Southeast Asian region during 2004.

The fourth article "**Coatings: Tiny Titans**" by *Larry Adams* explains that the nanotechnology revolution that is affecting so many industries is also making a huge impact on the coatings segment, as the addition of nano-engineered particles is increasingly used to enhance coatings performance to levels previously unimagined. Furthermore, the nanoparticles going into coatings are reaching ever-smaller sizes, and with the reduction in size comes an increase in capabilities and opportunities to use materials in new ways. The author further explains that part of a new breed of smart coatings, nanotech coatings can make a product harder, even at extreme temperatures, more scratch and abrasion resistant, less likely to crack, and offer improved thermal, optical and electrical properties, all the while keeping its aesthetic requirements. The article also highlights that we are in the early stages of a profound industry change. Using nano-based materials could reduce the cost of applying a coating from a few dimes per article down to 1 cent per article or less.

The subsequent article "**BIPHOR – Nanotechnology: For Waterborne Paint Improvement**", authored by *Fernando Galembeck, Maria do Carmo, V M da Silva, Renato Rosseto, Gilmar O Pinheiro* and *João de Brito* explains that BIPHOR is a new product for the paint industry based on contemporary nanotechnology concepts that are now reaching the semi-industrial production stage. It performs an outstanding combination of desirable functions in a waterborne paint that are described and explained in this article. The authors further explain that BIPHOR is a new family of Aluminium phosphates or polyphosphates made by a wet chemistry process. It is a "green chemistry," zero-effluent product made under mild temperature and pressure conditions that do not create any environmental problems during the fabrication process. Due to its chemical nature, BIPHOR residues in the paint industry or in the final user location may be safely discarded in the environment as a fertilizer component. It is a product as slurry as well as a dry powder. In both cases it is easily dispersed in water forming stable dispersions that have stable rheological properties. The article highlights that BIPHOR in waterborne paints leads to improved visual performance and durability at a reduced cost.

Jan H Schut, in the article "**Nanocomposites**" states that for skeptics who may doubt that nanocomposites have yet proven themselves commercially viable materials ranging from auto parts and precision moldings to wire and cable jacketing, there is plenty of evidence that 'nanos" are beginning to live up to their promise. These applications demonstrate that just a pinch of these tiny particles can cut weight and cost compared with higher loadings of conventional fillers. The benefits include improved mechanical properties, scratch resistance, barrier properties, fire resistance, and dimensional stability, plus faster extrusion and molding. The author explains that early applications have been motivated by cost savings relative to conventional glass and talc-filled resins. Nanomaterials weigh less, mold more easily on smaller presses, process dramatically faster and produce better properties. The most interesting potential application for nanoclays is thin-walling for lightweight parts. So far, people have used nanoclays to save money, but they haven't taken full advantage of nanotechnology to reduce wall thickness. Since nanocomposites are stiffer but lighter than glass or mineral-filled plastics and also enhance flow, they can reduce part weight and thickness. The article highlights that they are out of the lab and into the real world. In applications from automotive to agriculture, materials handling and electrical cable, nanoclay blends are lowering costs and raising performance. The market is already recognizing how nano-materials enhance performance.

Nanobiotechnology represents the convergence of nanotechnology and biotechnology, yielding materials and products that use biological molecules in their construction or are designed to affect biological systems. The article "**Nanobiotechnology in North Carolina**" authored by *Sarah Jackson, Maria Rapoza, Rudy Juliano, Kenneth Gonsalves* and *Ken Tindall* explains that the nanotechnology field is sufficiently developed and that discoveries are beginning to transition from scientific novelties to commercial applications. Nanotechnology-based products are already in the market in cosmetics and sunscreens, machinery coatings and electronic displays, and new applications may impact medicine, pharmaceuticals, genomics and proteomics in the near future. The next wave of applications is expected to include many from the nanobiotechnology field. The article further explains that applications of nanobiotechnology include: engineering biomolecules for non-biological use, such as DNA-based computer circuits using nanotechnology tools such as medical diagnostic devices and medical imaging to study biology or combining nanomaterials with biological systems for outcomes such as targeted drug therapies. It emphasizes that nanobiotechnology may have broad impacts across many sectors that are well-established including health care, textiles and electronics. This report describes nanotechnology, nanobiotechnology and some of their applications, with a focus on North Carolina's leadership role in nanobiotechnology today and in the years to come.

In the eighth article "**Nanotechnology Applications in Cancer**" the authors *Shuming Nie, Yun Xing, Gloria J Kim* and *Jonathan W Simons* explain that cancer nanotechnology is a cross-disciplinary field of research in science, engineering and medicine for cancer imaging, molecular diagnosis and targeted therapy. In this article, the current status of cancer nanotechnology and its applications in molecular tumor imaging, diagnosis and targeted therapy is discussed. Together with cancer bioinformatics and biocomputing, nanotechnology is one of the most promising and enabling technologies and it should make a significant impact in clinical oncology.

In the ninth article "**Change Required for the National Nanotechnology Initiative as Commercialization Eclipses Discovery**" the author *Matthew M Nordan* states that in the last seven years, emerging nanotechnology has increasingly become a fact of life and of business, as the technology has shifted from an era of discovery to one of commercialization. In this fashion, nanotechnology follows the example of other world-changing technologies like polymer science and biotechnology. The value chain starts with nanomaterials like carbon nanotubes and dendrimers, which are incorporated into intermediate products like memory chips and drug delivery systems, which are in turn used to make enhanced final goods like mobile phones and cancer therapies. Introduced in 2001 and signed into law in 2003, the US National Nanotechnology Initiative is the federal government's coordinating program for publicly-funded nanotechnology research, which has inspired similar efforts in countries worldwide from Germany to South Korea. Since the start of international competitiveness rankings in 2005, the position of the US has remained static while other countries have vaulted upwards in their nanotechnology activity. For example, nanotech funding is growing in the EU at twice the rate as in the United States, putting the EU on track to claim the mantle of nanotechnology leadership with a renewed focus on nanoscale science and engineering. Russia recently funded a state nanotechnology corporation with $5 billion of public financing. And scientists in China published nearly as many scientific journal articles on nanoscale science and engineering in 2007 as those in the US did. Emerging nanotechnology applications will affect nearly every type of manufactured product through the middle of the next decade, thereby getting incorporated into global manufacturing output.

In the next article **"Nanotechnology for Development or Knowledge Enclaves? The World Bank Case for Latin America"** the authors *Guillermo Foladori, Mark Rushton* and *Edgar Zayago Lau* state that a great deal has been written in recent years about nanotechnology and its revolutionary significance for science and real-world applications that are touted as being capable of profoundly transforming the world in which we live. Yet very little has been written about how they are incorporated into the context of the knowledge economy. In this article, the authors analyze the World Bank's intention to develop Scientific Millenium Initiatives as Centers of Excellence in Latin America to boost competitiveness and encourage economic growth, which is understood by the World Bank as a requirement for development. Nanotechnology is a strategic area within these projects.

In the eleventh article **"Nanotechnology: New Vistas for US"** the author *D Gayatri* says that according to National Science Foundation (NSF) of the US, the global market for nano product was estimated at around $1 trillion by 2015 and about $7.6 billion by 2003. The industry was expected to offer around seven million jobs during the next 10-15 years. The author further states that all the possible benefits that the technology could bring for the economy were encouraging the state and federal government to invest in research activities and infrastructure for building nanotech hubs. With Government support many regional nanohubs emerged. In addition, venture capitalists also seemed to be interested in this field and invested around 2225 million in the first quarter of 2003. The article highlights that the applications of nanotechnology ranged from imaginable ones of making things stronger and lighter by working precisely on each atom at a time and eliminating waste, to unimaginable applications such as using carbon nanotubes that were 100 times stronger than steel and only one sixth of their weight, to build space elevators. Identifying the importance of the new technology for the country the Government established the National Nanotechnology Initiatives (NNI) in 2000, which coordinated the activities in this field. The nanotech bill signed by President George Bush Jr., is a vital catalyst for the development and growth of what will become a $ 1trillion piece of the global economy. This bill will also allow the US to continue to strive for global leadership in the highly competitive nanotechnology marketplace. The bill according to analysts is a model of how the companies, Government and the universities could work together towards the growth of the economy.

In the succeeding article "**Societal Implications of Nanoscience and Nanotechnology in Developing Countries**" the authors *Birgit R Bürgi* and *T Pradeep* explain that nanotechnology, unlike any other technology, can find applications in virtually all areas of human life. In spite of being an infant at its evolution, some of the known issues related to nanotechnology suggest a wide spectrum of potential societal impacts. The current public nano-discourse provides sociology with a unique opportunity to switch from a merely passive, observational role to an active participating one, especially where the key players involved meet to find joint and concerted solutions for development. The objective of this article is to address the wider societal implications of nanoscience and its deriving technologies on society within the context of the developing world.

The authors *Fabio Salamanca-Buentello, Deepa L Persad, Erin B Court, Douglas K Martin, Abdallah S Daar* and *Peter A Singer* in the article "**Nanotechnology and the Developing World**" state that many developing countries have launched nanotechnology initiatives in order to strengthen their capacity and sustain economic growth. However, there has been no systematic prioritization of applications of nanotechnology targeted toward the challenges faced by the 5 billion people living in the developing world. The article shows that developing countries are already harnessing nanotechnology to address some of their most pressing needs. It identifies and ranks the ten applications of nanotechnology most likely to benefit developing countries, and demonstrates that these applications can contribute to the attainment of the United Nations Millennium Development Goals (MDGs). The article outlines a way for the international community to accelerate the use of these top nanotechnologies by less industrialized countries to meet critical sustainable development challenges.

The fourteenth article "**Micro Nano Technology Commercialisation Pitfalls**" by *Henne van Heeren, Patric Salomon, Lia Paschalidou* and *Ayman El-Fatatry* states that, ever since the realisation that MNT (Micro Nano Technology) offers potential opportunities for accessing new markets and expanding existing ones, the race for exploitation began. These opportunities encompassed technical, financial as well as enabling capabilities. As a result, successful ventures were initiated by (large) companies, institutes and entrepreneurs. These have charted the way forward. Failed adventures have, however, provided sufficient warnings of the pitfalls and challenges encountered along the way to commercialisation. This "white paper" attempts to summarise some of the more obvious problems and other less clear hurdles likely to be encountered by prospective new business ventures. Finally, in a time marked by many examples of paradigm shifts, such as that created by the micro, and even more so, the nano technology, new ventures get new opportunities and a chance of entering new markets.

The final article "**Will Nanotechnology Help Developing Economies**" by *Anil Varma* takes a look at issues in nanotechnology with regard to developing countries. A very relevant question it raises is "Will nanotechnology offer solutions to the problems faced by developing countries?" as a sequel to the previous question, "Will the new technology complement or substitute existing technologies for the similar problems?" If the technology is a substitute, then it raises the issue of possible adverse impact on employment. The article highlights that the important issue is access to developing countries in light of intellectual property rights. If this is not granted, it would mean most of the promises and advantages will be available to a select few in the world.

Nanotechnology is one of the most significant research areas which emerged in the past decade or so. It is based on the concept of creating applications based on components built on a very small (nano-) scale. Although a relatively new field, the impact of nanotechnology is already being felt. Indeed, many believe that as nanotechnology matures as a technology and an increasing number of applications become commercially viable, it will fundamentally alter how societies function. The field draws on knowledge and expertise from many (if not all) science and engineering disciplines, integrating aspects of physics, chemistry, biology, material sciences and biological and chemical engineering. Applications span the spectrum stretching from mature products such as sunscreens, emerging ones such as bio- or chemical sensors, to imaginative and yet-to-be conceived products such as self-replicating microscopic robot systems, which are currently in the realms of scientific fiction.

Section I

Standard Applications

1

Nanotechnology for Agriculture and Food Systems – A View

H C Warad and J Dutta

Nanotechnology has the potential to revolutionize the scientific world by allowing scientists to manipulate matter at the atomic or molecular scale using physics, engineering, chemistry and biology. Nanotechnology could be used to revolutionize the food and agriculture systems. Novel agricultural and food security systems, cellular biology, environmental protections, disease treatment delivery methods etc., are just a few examples where nanotechnology could have important impact. The article highlights that in agriculture and food systems, the fundamental life processes are explored through research in molecular and cellular biology. New tool for molecular and cellular biology are needed that are specifically designed for separation, identification and quantification of individual molecules. This is possible with nanotechnology. Nanotechnology applications in food and agriculture is in its nascent stage and over the next decade we will see increasing uses of tools and techniques developed by nanotechnology to detect carcinogenic pathogens and biosensors for improved and contamination-free food and agricultural products.

Introduction

The word 'nano' meaning 'dwarf' in Greek language refers to dimensions on the order of magnitude of 10^{-9}. Nanotechnology, focusing on special properties of materials emerging from nanometer size, for e.g., in biological systems, the first level of organization occurs at the nanoscale structure where all the fundamental properties and functions are systematically defined.

Nanotechnology has the potential to revolutionize the scientific world by allowing scientists to manipulate matter at the atomic or molecular scale using physics, engineering, chemistry and biology [1]. Nanotechnology is a broad and interdisciplinary area of research and development activity that has been growing at a rapid pace worldwide in the past few years. It enables researchers to understand the relationship between macroscopic properties and molecular structure in biological materials of plants and animal origin [2]. It is already having a significant commercial impact, which will certainly increase in the future. Using nanotechnology, scientists are able to self-assemble atoms into structures with highly controlled properties, e.g., nanowires [3, 4], self-assembled molecules and particles [5], 3-D architecture [6], etc.

There are two basic forms of attaining nanomaterials, "top-down" and "bottom-up". The Greeks coined the term 'atom', defined as the unbreakable, implying that they believed that, it was possible to break down matter to the level of individual atoms or molecules. This old-age approach is termed as 'top-down' approach. This approach usually involves breaking of big chunks of materials (physically or chemically) into smaller objects of desired shapes and sizes by either cutting or grinding e.g., mechanical milling, ion implantation, etc. Complementary to 'top-down' approach is the 'bottom-up" approach or 'self-assembly", which involves building up of macro-sized complex systems by combining simple atomic level components material. By this approach of arranging molecules one at a time, we can design complex systems by incorporating specific features which require a good understanding of individual molecular structures and various molecular forces.

From the moment of the creation of the universe to the first signs of life on this earth, self-organization has existed. Nature has evolved many bio-organic molecules that form complex structures with very complex dynamic behavior, called living cells. These cells, self-assemble and form further complex structures culminating in intelligent life forms like humans and other animals. Even when a damage is done to the living cells, nature has an amazing ability to heal itself by self-organization, e.g., when a living cell is wounded, the body reacts by sending white blood cells to ward off the infections killing the germs, red blood cells and proteins form a seal cover over the wound and also nutrients to the cells, so that they can produce new cells to replace the damaged cells. Such biomaterials are custom made for specific applications inspiring us to design materials that are ideal for a specific application rather than to cut and trim natural materials to suit our needs. When materials are built by the bottom-up process, one molecule at a time is possible to incorporate specific features at will. The concept of self-assembly with nanotechnology has a potential to impact diverse fields ranging from biology to materials science [7].

Nanotechnology will enable making high-quality products at a very low cost and at a very fast pace. It is commonly referred to as a generic technology that offers better-built, safer, long lasting, cheap and smart products that will find wide applications in household, communications, medicine as also agriculture and food industry amongst others. Currently the main thrust of research in nanotechnology focuses on applications like electronics, automation, medicine and life sciences [8]. The experience gained from this could be used to revolutionize the food and agriculture systems. Novel agricultural and food security systems, cellular biology, environmental protections, disease treatment delivery methods etc., are just a few examples where nanotechnology could have important impact.

Agriculture forms the backbone of Third World economies. Research in agriculture has always dealt with improving the efficiency of crop production, food processing, food safety and environmental consequences of food production, storage and distribution. Nanotechnology provides a new tool to pursue these historically relevant goals. A reappraisal of attitudes towards food security, increasing emphasis on environmental compliance and alterations to the Common Agricultural Policy all combine to make farming and rural land-owning far less profitable than the previous times.

In agriculture and food systems, the fundamental life processes are explored through research in molecular and cellular biology. New tools for molecular and cellular biology are needed that are specifically designed for separation, identification and quantification of individual molecules. This is possible with nanotechnology and could permit rapid advances in agricultural research, such as reproductive science and technology, conversion of agricultural and food wastes to energy and other useful by-products through enzymatic nanobioprocessing, disease prevention and treatment in plants and animals. To excel in these and other areas of agriculture, are required novel tools that allow us to work and explore living cells and biomolecules at the molecule scale. Nanotechnology holds such a promise.

Applications

Photocatalysis is one such application using nanoparticles [9]: Photocatalysis is a reaction in which chemical compounds react in the presence of light and itself not being completely consumed in the reaction. In the presence of UV light, the valance electrons in the nanoparticles are excited to form electron-hole pairs. These negative electrons and positive holes are strong oxidizers. When harmful substances (pesticides) stick to the positive holes, they are disintegrated into harmless compounds. The excited electrons are also injected in bacteria in contact of nanoparticles and hence act as a disinfectant (could find applications in fruit packaging and food engineering).

Photocatalysis degradation process has gained popularity in the area of wastewater treatment process [10]. Peral *et al.* [11] explained the use of photocatalysis for purification, decontamination and deodorization of air. Mills *et al.* [12] also explained semiconductor sensitized photosynthetic and photocatalytic processes for the removal of organics, destruction of cancer cells, bacteria and viruses.

Metal oxides like TiO_2 [13], ZnO [14], SnO_2 [15], etc., as well as sulphides like ZnS [16] have been used for photocatalysis. These nanoparticles have efficient disinfectant rate due to another important property of nanoparticles in general, which is increased surface to volume ratio (Figure 1). The principle of photocatalysis could be used in the decomposition of toxic pesticides, which take a long time to degrade under normal conditions [17].

Identification tags are a part of our daily life today, right from application in wholesale agriculture and live stock products to consumer products. Ultra miniaturized identification tags have applications in the field ranging from advanced biotechnology to agricultural encoding. The possibility of large number of combinations of these nanobarcodes (>1 million) makes them

Figure 1: Logarithmic Length, A Matter of Scale, and the Level of Structure

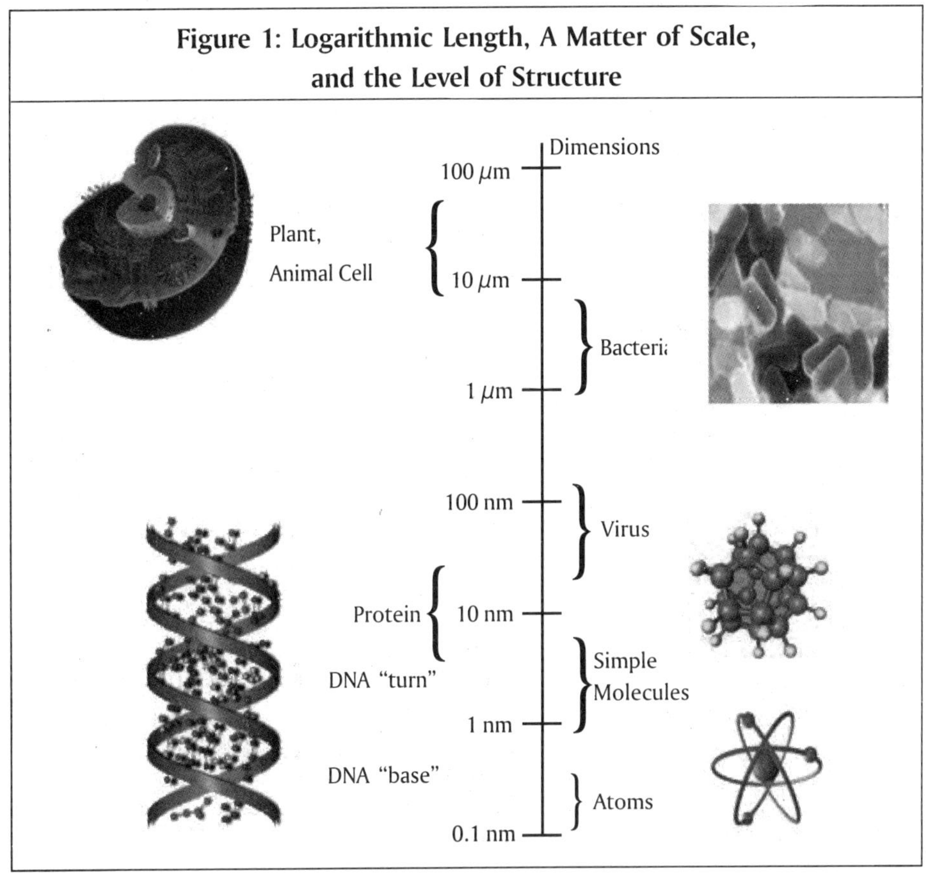

attractive also for use in multiplexed bioassays and general encoding. Matthew *et al.* [19] describes micrometer-sized glass barcodes doped with rare earth containing a specific type of pattern of different fluorescent materials that are identified by using UV lamp and optical microscope for application in DNA hybridization assays.

Nanobarcode [20] particles are encodable, machine-readable, durable, sub-micron sized taggants as shown in (Figure 3a). They are freestanding, cylindrically shaped metal nanoparticles having dimensions of 20-500 nm in diameter and 0.04-15 mm in length. The particles are manufactured in a semi-automated, highly scalable process by electroplating inert metals (Gold, Silver etc.) into templates defining particle diameter, and then releasing the resulting striped nano-rods from the templates (Figure 3b).

Figure 2: Number of Surface Atoms Increases when Particle Sizes Decrease. This Result in Increased Reactivity and other Physical and Chemical Properties Related to Exposure to Specific Conditions, Like Photocatalysis, Photoluminescence, etc. [18]

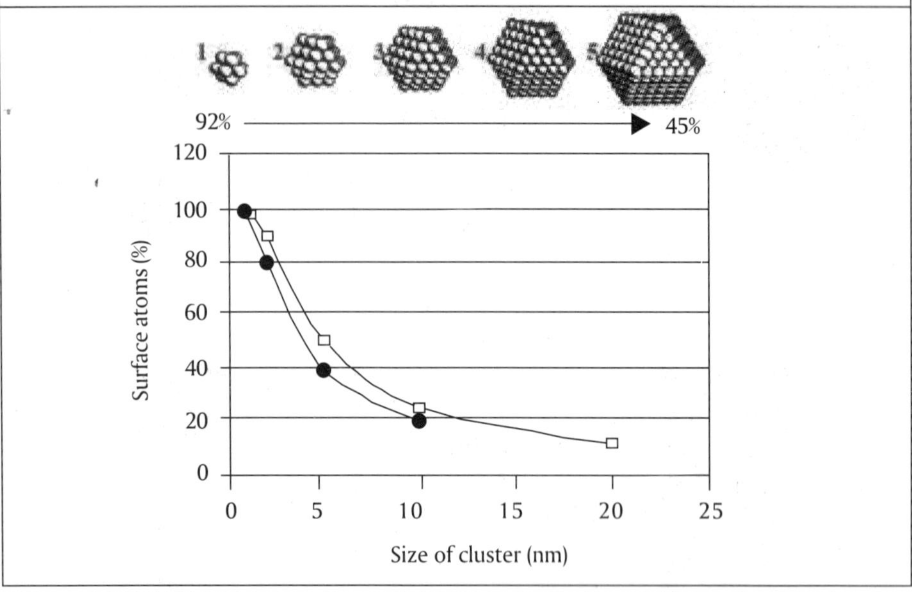

Figure 3a: Optical Micrograph of a Mixture of 7 Flavors of Nanobarcodes

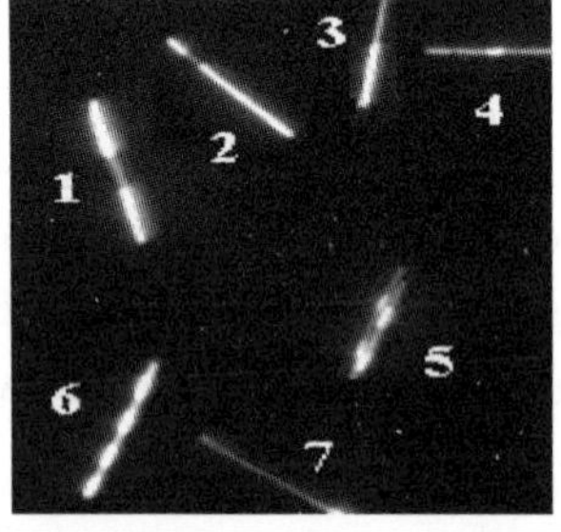

Figure 3b: Template-Directed Synthesis of Nanobarcodes Particles

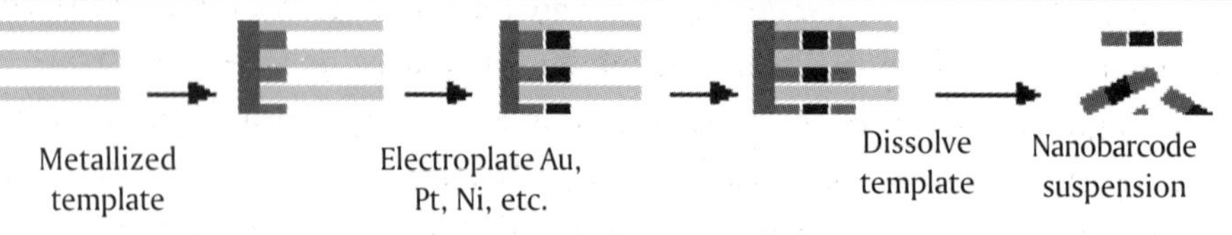

Source: http://www.nanoplextech.com/technology/nanobarcodes.htm

Applications for these nanobarcodes are as ID tags for multiplexed analysis of gene expression and intracellular histopathology. The Nanobarcodes particles technology also holds great promise in non-biological applications, especially for robust, uniquely identifiable nanoscale tagging of small items for authentication or tracking in agricultural food and husbandary products. The technology will allow tagging of items previously not practical to tag with conventional barcodes, as well as aiding in the development of new Auto-ID technologies.

Bacteria are the most primitive life forms found almost in all places. We come into contact with millions of bacteria every day. They are in the air we breathe, in the food we eat and on the surfaces of most things we touch. Along with some useful bacteria there are numerous other disease-causing bacteria. Organic dyes are the most commonly used biolabels to stain bacteria for detection. Organic dyes are expensive and also their fluorescence degrades with time. So there is a need for a better alternative. Recent advances in the field of luminescent nanocrystals have led to a new area of research in fluorescent labeling by Quantum Dots (QDs) with bio-recognition molecules. QDs have several prominent advantages over conventional organic fluorophores (dyes) as they are more efficient in luminescence compared to the organic dyes, their emission spectra are narrow, symmetric and tunable according to the particle sizes and material composition of the QDs, and they show excellent photostability [21]. Due to their broad absorption spectra, they can be excited to all colors of the QDs by a single excitation light source. Figure 4 shows bacteria bacillus bio-labeled by $ZnS:Mn^{2+}$ nanoparticles capped with bio-compatible 'chitosan' [22].

Figure 4: Bacillus as Seen under a Fluorescence Microscope – The Orange Glow is from Bio-Functionalized $ZnS:Mn^{2+}$[24]

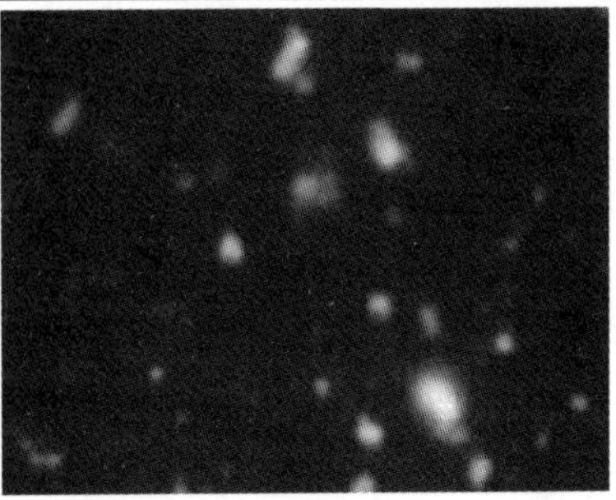

Su *et al.,* [23] demonstrated a sensitive and rapid method for the detection of E. Coli O157:H7 using Quantum Dots (QDs) as a fluorescence marker coupled with immunomagnetic separation. They used magnetic beads coated with anti-E. Coli O157 antibodies to selectively attach target bacteria, and biotin-conjugated anti-E. Coli antibodies to form sandwich immuno complexes. After magnetic separation, the immuno complexes were labeled with QDs via biotin-streptavidin conjugation.

Pathogen detection market encompasses medical, military, food and environmental industries. The food pathogen testing markets are close to 1 billion US$ [25]. Micro-organisms produce a range of characteristic volatile compounds that may be useful as well as harmful to human beings; e.g., Yeasts are beneficial for fermentation, bacteria eat sugar thereby, producing alcohol as a by-product. Dairy products, bakery products and other food products are ideal media for rapid growth for a wide range of micro-organisms. Bacteria are the most common cause of food-rotting. The presence of foul odor is an indication for food rotting. Human nose can literally detect and distinguish thousands of odors, which is sometimes impractical and could also be a further cause for poisoning. Sometimes, it is more practical to use instruments to detect these odors with what we popularly known as rapid detection biosensors which can minimize the need for food processors to perform lengthy microbial testing and immunoassays on materials suspected of carrying food-borne pathogens.

Applications include detecting contamination in water supplies, raw food materials and food products as well as the processing lines – all of which require food producers to either hold on to inventory to complete the tests or simply release products which might be harmful. Enzymes can be used as a sensing element, since they are known to be very specific in attachment to certain biomolecules. The general classifications and working of enzymic biosensors is described in Figure 5.

Electronic Nose (E-Nose) is a device mimicking the operation of the human nose, which uses a pattern of response across an array of gas sensors to identify different types of odors. The main purpose of the E-Nose is to identify the odorant, estimate the concentration of the odorant and find characteristic properties of the odor as might be perceived by the human nose.

Figure 5: Presence of the Biological Element is Determined through Electrical Means (Amperometry or Potentiometry) or through Optical Means [26]

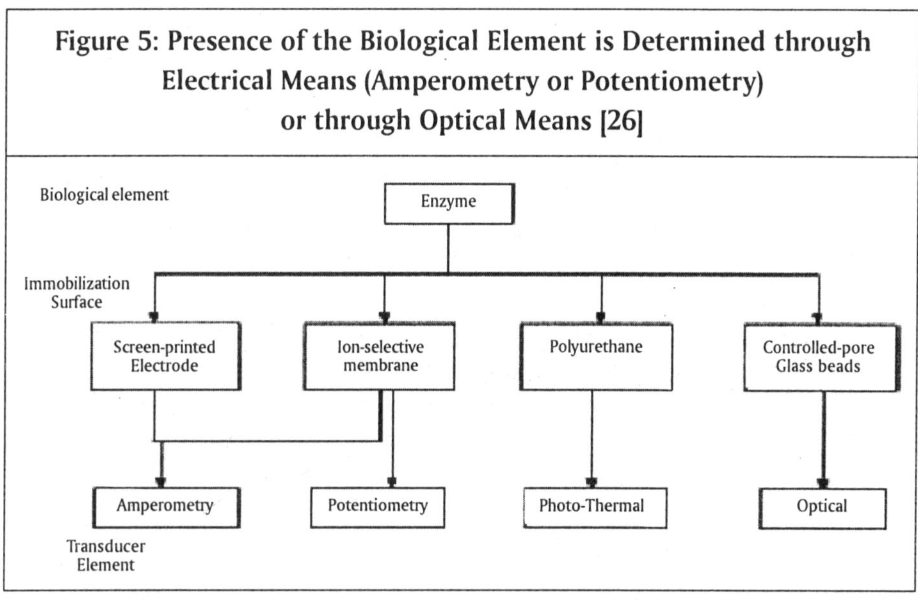

The main components in an E-Nose are its gas sensors that identify odors. One such gas sensor functional diagram is shown in Figure 6a. These gas sensors are composed of nanoparticles (e.g., zinc oxide nanowires [4, 27]) whose resistance changes when a certain gas is made to pass over it. This change in resistance generates

Figure 6a: Functional Diagram of Gas Sensors – Whose Current is a Measure of Change in Resistivity of ZnO Nanowires on Gas Adsorption [27]

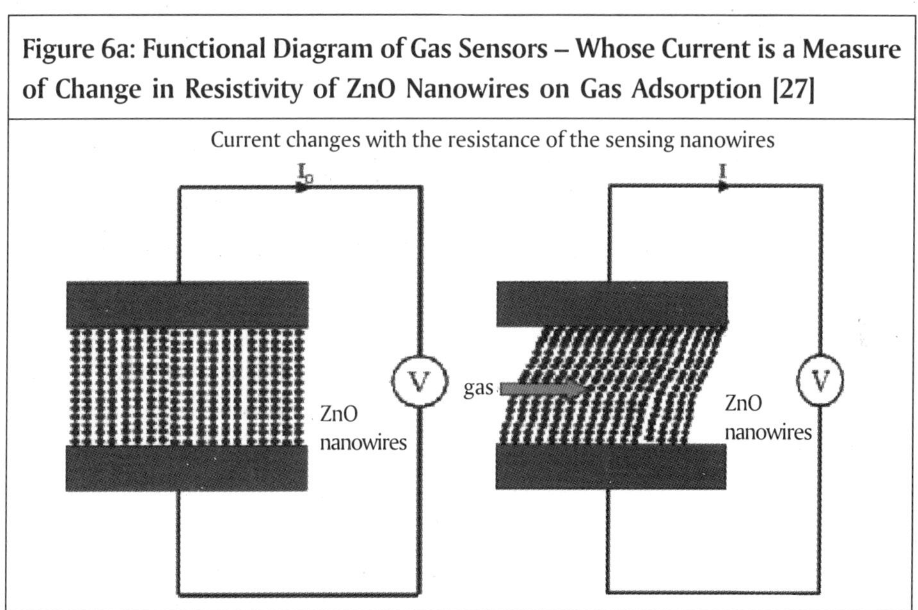

a change in electrical signal that forms the fingerprint for gas detection. This fingerprint pattern derived from the sensor is used to determing the type, quality and quantity of the odor being detected. The advantage of using nanoparticles is that they have improved surface area for better gas adsorption (Figure 2). Figure 6b shows Scanning Electron Microscope (SEM) image of ZnO nanowires.

Figure 6b: SEM Image of Homogenous ZnO Nanowires

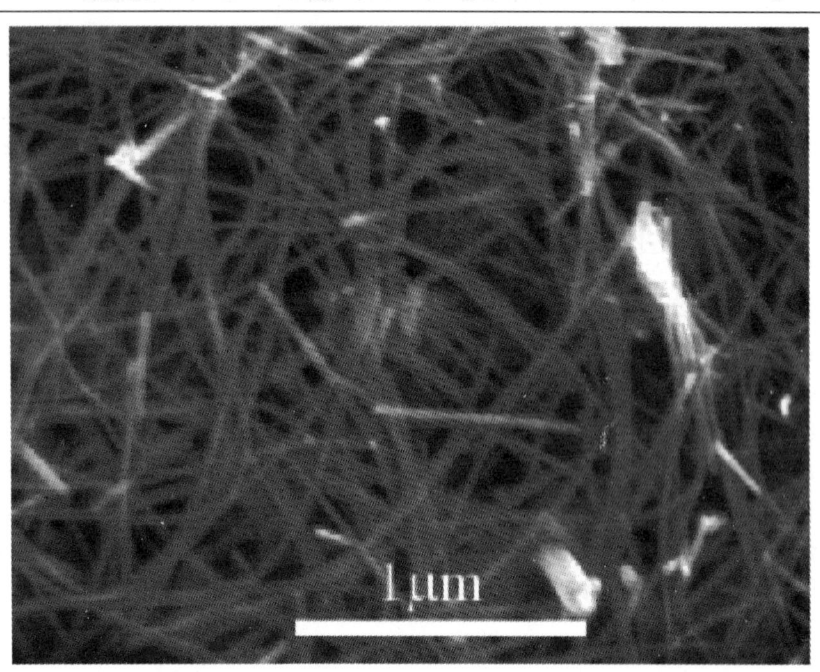

Gold has fascinated man for centuries. Its several qualities have made it valuable throughout history, both as a medium of exchange and for decorative use as jewelry. Like bulk gold, it is most widely studied and abundantly used nanoparticles. Currently rapid testing arrays like pregnancy tests and bio-molecule detectors are a few commercial applications of gold nanoparticles. These applications are based on the fact that the color of these colloids depends on the particle size, shape, Refractive Index (RI) of the surrounding media and separation between the nanoparticles. Any change in any of these parameters results in a quantifiable shift in the Surface Plasmon Response (SPR) absorption peak [28].

By carefully choosing the capping agent for stabilizing gold nanoparticles, we can make these nanoparticles attach to specific molecules, which get adsorbed on the surface of these nanoparticles changes the effective RI of the immediate surrounding of the nanoparticles [30] (Figure 7a). If the detecting molecules (bio-macromolecules) are larger than the gold nanoparticles, they will absorb a few nanoparticles making them agglomerate into lumps [31]. This reduces the particle spacing, resulting in shift of SPR, and hence changing the color of gold nanoparticles (Figure 7b).

Figure 7: Plasmon Resonance Sensors – (a) The changes in local RI caused by adsorption of amine molecules, causes a detectable shift in the SPR peak; (b) Agglomeration of gold nanoparticles causes a red shift in the SPR peak, agglomeration can be induced with a macro-molecule [29]

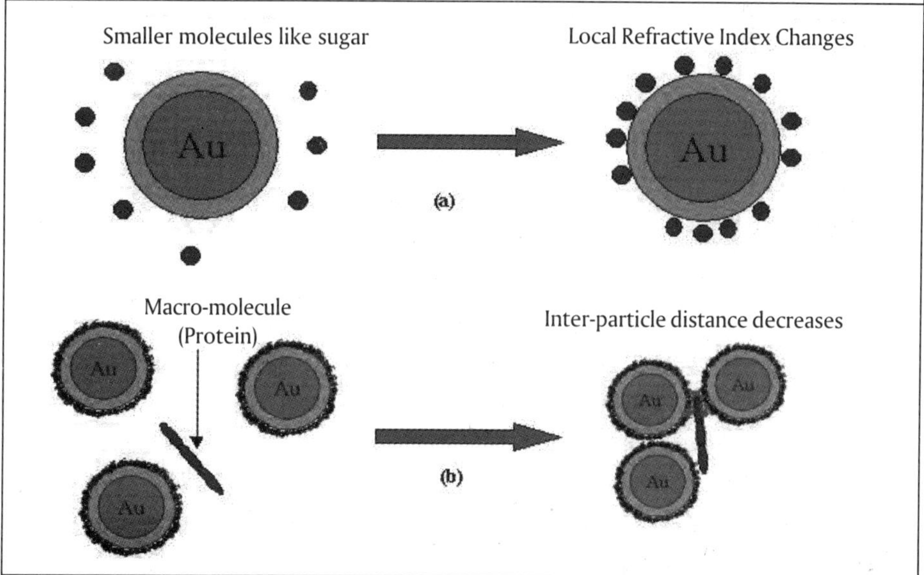

Conclusions and Perspectives

Nanotechnology is a part of any nation's future. Research in nanotechnology has extremely high potential to benefit society through applications in agriculture and food systems. Any new technology carries an ethical responsibility for wise application and the recognition that there are potential unforeseen risks that may come with the tremendous positive potential. As a part of a nation's future, it is critical that the future workforce be trained in nanotechnology. The first step is informing the public at large about the advantages and challenges of nanotechnology. As public awareness increases, so will interest in the understanding of nanotechnology and new applications in all the domains will be found. Rapid testing technologies and biosensor related to the control of pests and cross contamination of agricultural and food products will certainly see applications of nanotechnology in the very near future.

As in the case of almost every nonconventional technology, e.g., genetic engineering, some fear that nanotechnology can give people too much control. We believe that this control can be wisely used, and that the huge contributions that nanotechnology can make are very strong arguments in favor of using this revolutionary science to its fullest potential. Food and agriculture technology should take advantage of the powerful tools of nanotechnology, for the benefit of humankind.

Richard Feynman, an eminent physicist said in a famous lecture in 1959 [32], "*The biological example of writing information on a small scale has inspired me to think of something that should be possible. Biology is not simply writing information; it is doing something about it. A biological system can be exceedingly small. Many of the cells are very tiny, but they are very active; they manufacture various substances; they walk around; they wiggle; and they do all kinds of marvelous things – all on a very small scale. Also, they store information. Consider the possibility that we too can make a thing very small which does what we want – that we can manufacture an object that maneuvers at that level!*" And that was 45 years ago! Of course it isn't yet possible to realize all that he predicted, or dreamt but the future will definitely show new applications of nanotechnology in the field of food and agriculture.

(H C Warad is a Laboratory Supervisor and Research Associate and J Dutta is group leader at Asian Institute of Technology.)

References

[1] Roco, M C, Williams, R S, Alivasatos, P, Eds. Nanotechnology Research Directions: IWGN Workshop Report, Kluwer Academic Publishers: Norwell, MA, 1999.

[2] F Kulzer and M Orrit, "Single-Molecule Optics", (2004), *Annual Review of Physical Chemistry*, Vol. 55, 585-611.

[3] M H Huang, Y Wu, H Feick, N Tran, E Weber and P Yang, "Catalytic Growth of Zinc Oxide Nanowires by Vapor Transport" (2001), *Adv. Mater.*, Vol. 13, 113-116.

[4] M K Hossain, S C Ghosh, Y Boontongkong C Thanachayanont and J Dutta, "Growth of zinc oxide nanowires and nanobelts for gas sensing applications", (2005), *Journal of Metastable and Nanocrystalline Materials*, Vol. 23, 27-30.

[5] J Dutta and H Hofmann, "Self-Organization of Colloidal Nanoparticles", (2004) *Encyclopedia of Nanoscience and Nanotechnology*, Vol. 9, 617-640.

[6] L Vayssieres, K Keis, S E Lindquist and A Hagfeldt, "Purpose-Built Anisotropic Metal Oxide material: 3D Highly Oriented Microrod Array of ZnO" (2001), *J Phys. chem.* B, 105, 3350-3352.

[7] P Liu, Y W Zhang and C Lu, "Finite element simulations of the self-organized growth of quantum dot superlattices", (2003), *Phys. Rev.* B 68, 195314.

[8] *http://www.nano.gov/*

[9] M Blake, "Bibliography of Work on Photocatalytic Removal of Hazardous Compounds from Water and Air", (1997), NREL/TP-430-22197, National Renewable Energy Laboratory, Golden.

[10] J M Herrmann, "Heterogeneous photocatalysis: fundamentals and applications to the removal of various types of aqueous pollutants, (1999), *Catalysis Today*, Vol. 53, 115–129.

[11] J Peral, X Domenech and D F Ollis, "Heterogeneous photocatalysis for purification, decontamination and deodorization of air", (1997), *J Chem. Technol. Biotechnol*, Vol. 70, 117-140.

[12] A Mills, L Punte, and M Stephan, "An overview of semiconductor photocatalysis", (1997), *J Photochem. Photobiol. A.*, Vol. 108, 1-35.

[13] D S Bhatkhande, V G Pangarkar and A A C M Beenackers, "Photocatalytic degradation for environmental applications – A Review", (2001), *J Chem. Technol. Biotechnol*, Vol. 77, 102-116.

[14] D Li, H Haneda, "Morphologies of zinc oxide particles and their effects on photocatalysis", (2003), *Chemosphere*, Vol. 51(2), 129-37.

[15] Y A Cao, X T Zhang, W S Yang, H Du, Y B Bai, T J Li, J N Yao, "Bicomponent TiO_2/SnO_2 particulate film for photocatalysis", (2002), *Chem. Mater.*, Vol. 12, 3445.

[16] L C Torres-Martínez, L Nguyen, R Kho, W Bae, K Bozhilov, V Klimov and R K Mehra, "Biomolecularly capped uniformly sized nanocrystalline materials: glutathione-capped ZnS nanocrystals", (1999), *Nanotechnology,* Vol. 10, 340-354.

[17] A Malato, J Blanco, A Vidal and C Richter, "Photocatalysis with solar energy at a pilot-plant scale: an overview", (2002), Applied Catalysis B: Environmental, Vol. 37, 1–15.

[18] *Nanomaterials* by J Dutta and H Hofmann, Text book in preparation.

[19] M J Dejneka, A Streltsov, S Pal, A G Frutos, C L Powell, K Yost, P K Yuen, U Muller, and J Lahiri, "Rare earth-doped glass microbarcodes", (2003), *PNAS,* Vol. 100, 389-393.

[20] S R Nicewarner-Pena, R G Freeman, B D Reiss, L He, D J Pena, I D Walton, R Cromer, C D Keating and M J Natan, "Submicrometer metallic barcodes" (2001), *Science,* Vol. 294 (5540), 137-41.

[21] R Vacassy, S M Scholz, , J Dutta, C J G Plummer, R Houriet and H Hofmann , "Synthesis of controlled spherical Zinc Sulfide Particles by Precipitation from Homogeneous Solutions", (1998), *Journal of American Ceramic Society* Vol. 81, 2699-2705.

[22] H C Warad, C Thanachayanont and J Dutta, "Highly Luminescent Manganese Doped ZnS Quantum Dots for Biological Labeling", SmartMat-'04, International Conference on Smart Materials, Smart/Intelligent Materials and Nanotechnology, Chiang Mai, Thailand, 1-3 December 2004, (Manuscript in Preparation).

[23] X L Su, Y Li, "Quantum dot biolabeling coupled with immunomagnetic separation for detection of Escherichia coli O157:H7" (2004), *Anal Chem.* Vol. 76 (16), 4806-10.

[24] H C Warad, S C Ghosh, C Thanachayanont and J Dutta, "Highly Luminescent Manganese Doped ZnS Quantum Dots for Biological Labeling", conference proceedings 'Smart/ Intelligent Materials and Nanotechnology', Chiang Mai, Thailand, December 1-3, 2004, (Manuscript in Preparation).

[25] E C Alocilja and S M Radke, "Market analysis of biosensors for food safety", *Biosensors and bielectronics,* (2003), Vol. 18, 841-846.

[26] P D Patel,"(Bio) sensors for measurement of analytes implicated in food safety: A review" (2002) *Trends in analytical chemistry,* Vol. 21, 96-115.

[27] A Sugunan, H C Warad, C Thanachayanont, J Dutta And H Hofmann, "Zinc oxide nanowires on non-epitaxial substrates from colloidal processing for gas sensing applications", Nanostructured and Advanced Materials for Applications in Sensor, Optoelectronic and Photovoltaic Technology, NATO-Advance Study Institute, Sozopol, Bulgaria, September 6-17, 2004.

[28] J C Valmalette, L Lemaire, G L Hornyak, J Dutta, and H Hofmann,"Optical properties of metal cluster based nanocomposite" (1996) *Analysis Magazine,* Vol. 24, 23-25.

[29] A Sugunan and J Dutta, "Nanoparticles for nanotechnology", *Magazine of Physics, Science and Idea,* (In Print).

[30] N Nath and A Chilkoti, "A colorimetric gold nanoparticle sensor to Interrogate Biomolecular Interactions in Real Time on a Surface" (2004), *Anal. Chem.*, Vol. 74, 504-509.

[31] G Carrot, J C Valmalette, C J G Plummer, S M Scholz, J Dutta, H Hofmann and H Hilborn, "Gold Nanoparticle Synthesis in Graft Copolymer Micelles", (1998), Vol. 276, 853-859.

[32] R P Feynman, "There's plenty of room at the bottom", (1960), Engineering and Science, *http://www.zyvex.com/nanotech/feynman.html*

Nanotechnology and Agriculture

– Anil Varma

Out of the current global population almost half of the population lives in Asia. A major section of those who are in developing nations on a daily basis face the situation of food shortage which is the effectively fallout of impacts of the environment as well as the instability in the political scenario, whereas, on the other hand, there is a surplus of food in the developing nations. In developing nations the concentration is on the development of crops that are resistant to droughts as well as pests.

The forecast is that the application of nanotechnology to the industries in the agricultural and food sector will comprehensively change the food industry by changing the manner in which food is produced, along with changing the processing, packaging, transportation and the manner of consumption. Nanotechnology has the prospective capacity to bring about a revolution in the agriculture and food industry with innovative tools for treating diseases molecularly, rapidly detecting diseases, developing the capacity of the plants for nutrient absorption etc. The agricultural industry will be able to effectively fight viruses. The near future will bring about the availability of nanostructured catalysts which will enable increased effectiveness of pesticides and herbicides which will allow the usage of lower doses. Nanotechnology will also enable the protection of the environment in an indirect manner through the usage of supplies of alternative or renewable energy, as well as the usage of filters or channels to decrease pollution and cleaning the pollutants that are in existence.

(Anil Varma, Consulting Editor, The Icfai Research Centre Pune. He can be reached at avarma@iupindia.org.)

2

Real Life Applications of Nanotechnology in Electronics

Alan Rae

Nanotechnology is like a toolkit for the electronics industry. It gives us tools that allows us to make nanomaterials with special properties modified by ultra-fine particle size, crystallinity, structure or surfaces. These will become commercially important when they give a cost and performance advantage over existing products or allow us to create new products.

Nanotechnology is receiving a lot of attention from companies, universities and governments. The US National Nanotechnology Initiative is matched by initiatives in Europe and Asia. But what does it mean for existing businesses and new businesses in the electronics market. Is it a real tool for today...or, are the applications way out in the future? Will it be economic or outrageously costly?

The presentation will outline areas in Nanotechnology with specific impact on semiconductors, passive components, display materials, packaging and interconnection.

The Promises

The collection of synthesis techniques collectively known as Nanotechnology presents many opportunities to reshape the electronics industry from top to bottom.

Nanotechnology can offer us:

- Uniform particles: metal, oxide, ceramics, composite;
- Reactive particles: as above;
- Unusual optical, thermal and electronic properties: phosphors, heat pipes, percolation based conductors;
- Nano-structured materials: tubes, balls, hooks, surfaces; and
- Self-assembly: liquid-based, vapor based or even by diffusion in the solid state.

The 2004 iNEMI roadmap is a comprehensive survey that reviews the issues affecting the electronics supply chain. Gaps in the technology or infrastructure that can adversely affect iNEMI members are identified, and the iNEMI Research Committee was formed to prioritise and the disposition of the tasks and to identify companies, universities and government laboratories that can address them for the mutual good.

Almost every roadmap chapter in 2004 identifies aspects of Nanotechnology that can enhance existing products or to replace their structure, function. Some of these are outlined below.

Long Term Issues

Once CMOS technology dips below about 20nm resolution, quantum effects such as electron tunneling start to result in phenomena like unacceptable leakage. The only way to move below that size is to utilise these and other quantum effects in new types of minute structures, be they pure electronic or bio-electronic (remember, the most effective and energy efficient computer available sits on your shoulders!). We know that if we extrapolate Moore's law we "hit the wall" with CMOS about 2015 and although we don't know which technology will replace it, it may well be disruptive. On the other hand, using atomic cluster deposition as described later, may allow us to extend much of the established silicon fabrication infrastructure while creating nanoscale structures.

Mid-Term Issues

In many areas of technology, once we hit an area of concern, we can develop a workaround. Hence, clock speed is the measure for processing capability, which has been followed by many, has been replaced in some devices by distributed processing with two processors placed on the same chip. This gets the job done without reducing heat penalty and gives us a breathing space, many upper end processors generate between 100W and 200W, but the heat issue has not gone away. Several unusual properties of nanoscale materials, enhanced thermal conductivity of Carbon nanotubes, diamond-like films, nano-metal dispersions, have the promise of aiding heat removal.

Shorter-Term Issues

Enhancement of shielding materials, solders, conductive adhesives, underfills etc., is now becoming possible as nano-sized materials become available and economically viable.

Technology Push and Market Pull – The Commercialisation Challenge

Many nano materials have been developed because of their interesting properties and companies have been founded on products for which there is limited market demand (see the left half of Figure 1). This tends to lead to leading edge products with very limited immediate commercial potential.

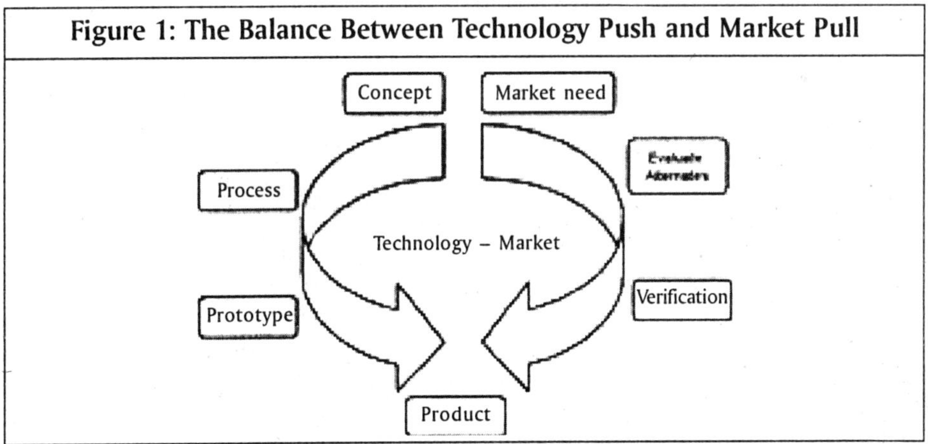

Figure 1: The Balance Between Technology Push and Market Pull

On the other hand, the approach of many established companies has been the Market Pull approach where existing solutions are sought for market needs (see right half of Figure 1).

This conservative approach can result in a very small increment in performance which may not show cost-benefit improvement for that particular application.

A more balanced approach followed by a number of successful companies, is to take a parallel track, constantly reviewing technology choices on a portfolio basis and applying them to market needs. Technology platforms thus developed, such as metal powders, diamond-like coatings, Carbon nanotubes or atomic cluster deposition, can be applied to several other business areas in addition to pure electronics, such as structural engineering, life sciences or energy.

Nanotechnology should only be applied where there is an economic advantage coupled to a performance advantage. This is seen in industrial processes as well as consumer goods where a "luxury" new technology becomes the standard once the existing technology is reaching its limits and the new technology starts to take hold. An example in the consumer field is VCR vs. DVD vs. DVD-R vs. DVR; in the industrial field embedded capacitors vs. discretes. In each case the shape of the graph may differ but the important point is that there is a "crossover" which marks the start of market adoption of the new product (Figure 2).

Figure 2: Replacement of an Existing Technology by a New Technology Based on Economics

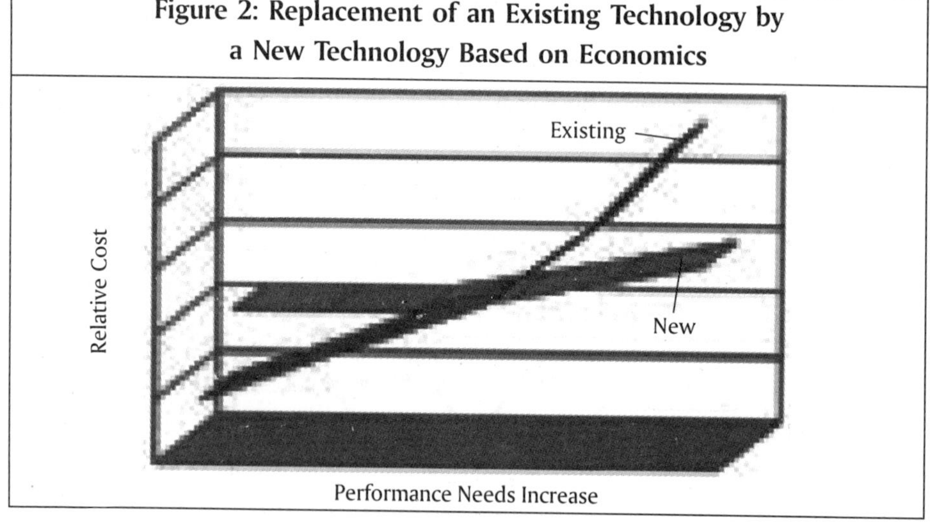

Nanotechnology Application Areas: Semiconductors

Some of the most revolutionary applications in nanotechnology are in the semiconductor areas. As the semiconductor roadmaps look out towards 2015 and below 20 nm features, the need for different structures is becoming apparent...once we move to ultraviolet and then Xray lithography, there is nowhere to go (in a practical sense) to image ultra small features.

Imagine doping a Carbon or Silicon nanotube, coating it with differently doped materials, assembling it (preferably self-assembling it) in an array. Imagine creating quantum dots that can store a single electron charge. Imagine trapping atoms inside a nanotube and using the electron spin to create a quantum computing device. There is a large number of potential routes to new computing, storage and optical devices. The devices we are making now are quite clumsy compared with established semiconductor technology. But they will surely improve!

One example of a semiconductor technology is generating great interest – the atomic cluster deposition technology pioneered by Nano Cluster Devices (NCD) of Christchurch, New Zealand.

The technology results from the convergence of two well-established technologies, atomic cluster deposition and the type of lithography currently used in semiconductor manufacture.

The technology revolves around an ability to fabricate nanoscale wirelike structures by the assembly of conducting nanoparticles. The attractions of this approach are:

- Electrically conducting nanowires can be formed using only simple and straightforward techniques, i.e., cluster deposition and relatively low resolution lithography;
- The resulting nanowires are automatically connected to electrical contacts;
- Electrical current can be passed along the nanowires from the moment of their formation;
- No manipulation of the clusters are required to form the nanowire because the wire is "self assembled" using one of two techniques described below; and
- The width of the nanowire can be controlled by the size of the cluster that is chosen.

One of the merits of this technique is that it is extremely simple: nanoscale particles, formed by inert gas aggregation are deposited from a molecular beam onto prefabricated lithographically defined nanocontacts. The cluster deposition is random but is managed via a novel templated surface strategy or within percolation theory. In the templating approach surface structures guide the particles to the 'correct' positions so as to form a wire. In the percolation approach, a deep understanding of the theory and sophisticated computer simulations have been used to design device geometries that ensure a single wire-like path whcih is formed between the contacts near the percolation threshold. In either case, an electrically conducting nanowire, which is automatically connected to electrical contacts, is therefore formed with no need to manipulate particles individually or use complex fabrication techniques. The width of the wire can be controlled by the size of the deposited particles.

Applications include:

- Chemical sensors, including Hydrogen and glucose sensors;
- Read heads for hard disk drives;
- Transistors, interconnects and integrated circuits (semiconducting and conducting wires);
- Photosensors;
- Deposition control systems, a spin-off technology for high precision control of particle deposition in the sub-monolayer regime.

Products are under development with several companies (Figure 3).

Figure 3: An 8x8 mm Square Chip Package (Low Resolution, Left) with Lithographically Defined Contacts (Higher Magnification, Centre) and a Cluster Assembled Wire (Highest Resolution, Right)

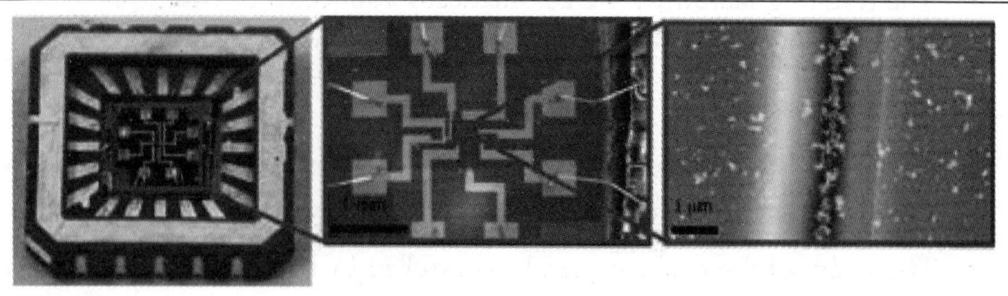

Packaging IC and MEMS Devices

Packaging of advanced devices is going to continue to be problematic since temperature is the enemy of ultra-fine features which can be easily destroyed by thermal diffusion or differential expansion. In the shorter term, improved fillers for mold compounds (which are already ~90% filler) and underfills can lead to better thermal and electrical performance combined with easier flow properties.

The use of high thermal conductivity Carbon nanotubes and diamondlike films with thermal conductivity over twice that of Copper will provide worthwhile solutions as will CTE matched fillers with conductive and dielectric properties.

Board/Substrate

In the last three years the board business has changed from a commodity business to a specialty business with materials optimised for thermal, high frequency and environmental reasons. Boards still need improvement in areas such as CTE and flatness and the embedding of passive components needs a low-cost self-assembly type process in order to lower the costs to make it truly competitive.

Improved ceramic substrates are possible, but in fact, many ceramic operations already use the principles of Nanotechnology in developing precursor particles with high reactivity and uniformity. There are options to improve conductor and embedded passive technology and to strengthen the substrates as an average substrate size is increasing due to the greater use of modules.

Passive Components

There are many uses of interconnection materials within passive components. Monosize materials hold promise in Tantalum capacitors and ceramic capacitors, where improved termination materials and electrode materials promise a further reduction in size and cost. In particular, reducing the electrode thickness in base metal ceramic capacitors promises a significant improvement in volumetric efficiency.

Advanced nano Copper and Nickel powders available in 2005 offer the following advantages (Figure 4):

Figure 4: 200 nm Ni Powder and 600 nm Cu Powder

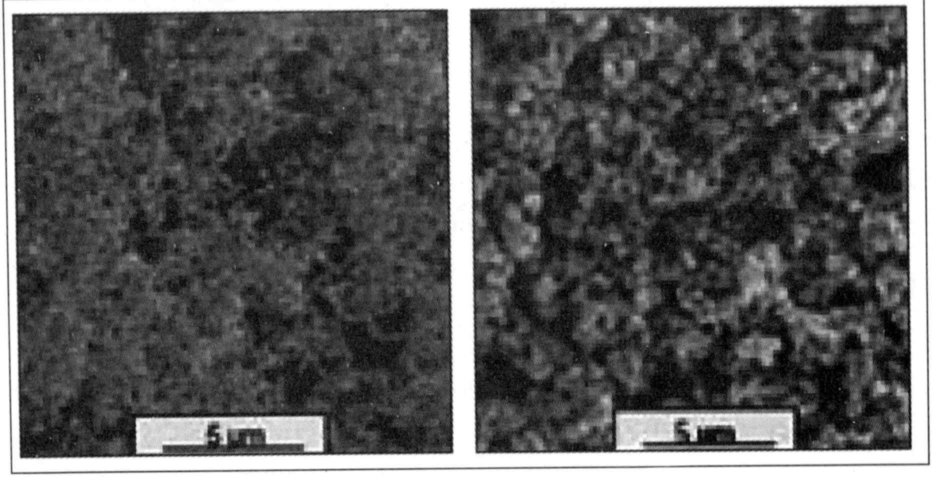

- Controllable particle size;
- Virtually monosize particle size distribution;
- Monolayer oxidation resistant coating;
- Economic and scalable process. Barium titanate and other dielectric materials are already produced by "wet processes" such as hydrothermal and oxalate precipitation that produce nano-sized barium titanate particles that are subsequently grown by thermal treatment to develop optimum properties. As we move down beyond dielectric thicknesses of 1 micron we will have to move from conventional tape casting to technologies such as gel casting or vapor phase technologies to optimise performance.

Displays

Display technology is similar to semiconductor, where the number of techniques to assemble devices is bewildering! Enhanced plasma displays using nanotubes, light emitting nanowires, quantum dot arrays, a lot of development work needs to be done but new products can be expected in the 2006-2007 time frame. Although Carbon nanotubes have been extremely expensive, thousands of Dollars a gram, new processes are dramatically reducing the cost of production and the cost will reduce past the cost of Silver flake within 5 years.

Consumer and Industrial Products

Finally, Nanotechnology can contribute in case design and EMI shielding. A major cell phone manufacturer has stated in public that they will have nanotube reinforced phone cases within 2 years, stronger, lower weight, EMI shielding and recyclable.

Power Delivery

A great deal of work is being done this area to enhance the properties of portable energy systems, enhancing the performance of Li-ion batteries and fuel cells.

Nano-sized electrolyte, anode and cathode materials have already increased the performance of solid oxide fuel cells (SOFC) while reducing the operating temperatures and start-up time dramatically (Figure 5). These will become commercially available in 2005.

Figure 5: Compact Solid Oxide Fuel Cell Using Nanotechnology in the Ceramic and Metal Structures

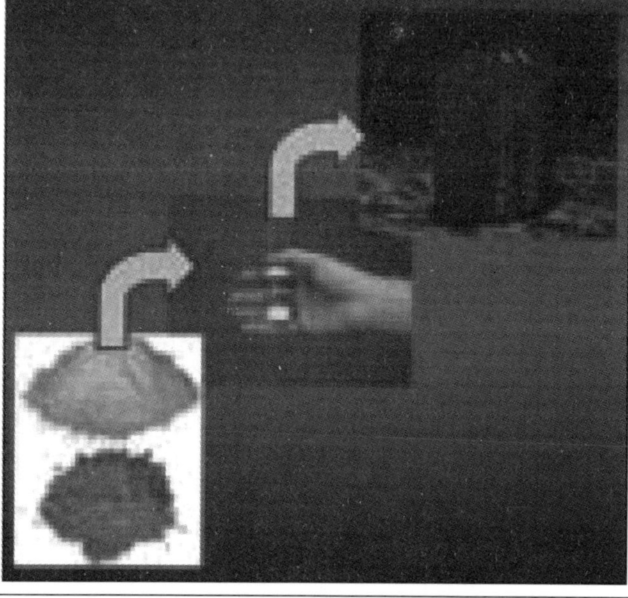

Conductive Adhesives

Most conventional isotropic adhesives use Silver flake in a resin base, typically epoxy, and have limitations in terms of conductivity, strength and moisture resistance. They typically cure in the 120-150°C range, much lower than the reflow temperature of solders.

Copper flakes with oxidation resistant (Figure 6) or Silver coatings are also being employed in conductive adhesive and EMI shielding.

Figure 6: High Aspect Ratio Oxidation Resistant Copper Flake, Thickness 150 nm

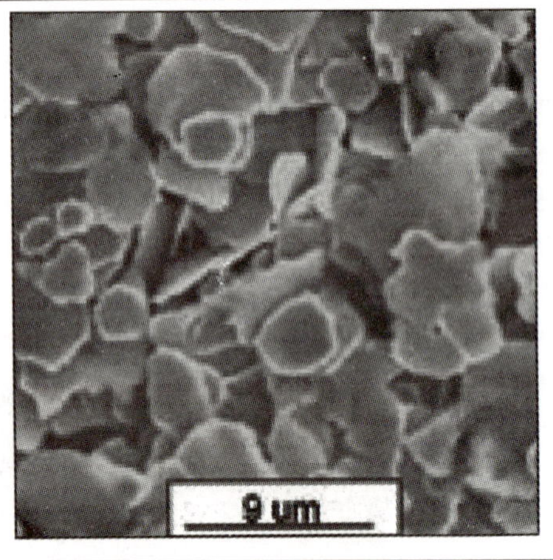

Anisotropic adhesives are largely based on Au plated polymer spheres dispersed in a resin. As the resin shrinks on cooling the conductive surfaces are pulled together and the spheres compressed between them.

There are many opportunities to use novel conductive adhesive systems to reduce processing temperatures but issues of strength under physical shock and high moisture environments still have to be solved. Nanomaterials of various types can exploit percolation and other effects to produce stronger, more conductive adhesives, be they polymer or frit based.

Novel Attachment Structures

Conductive hook, loop and other types of mechanical fasteners based on Carbon nanotubes have the potential of replacing conventional inter-connections as strong entangled loops promise thermal and electrical conductivity, compliance and strength, calculated at up to 30 times the strength of conventional adhesives on an area basis. Arrays of hooks can be synthesised which can entangle other hooks or a nanotube mat as shown in Figure 7.

Figure 7: Carbon Nanotube Fibre Mat and a Nanotube Hook Attachment Concept

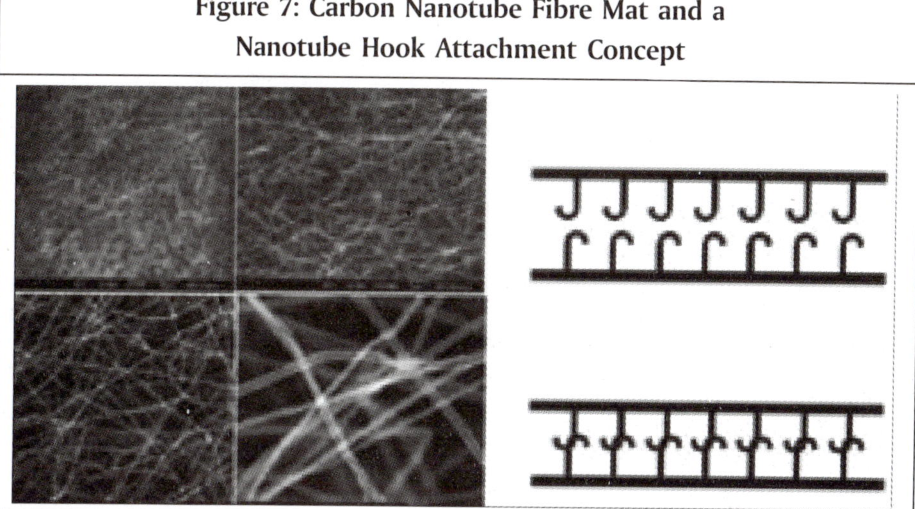

Summary

Over the next five years we will see significant introduction of nanomaterials and novel production processes based on Nanotechnology which will address key issues of importance to the electronics industry. The use of Nanotechnology, in long term, will allow us to meet customer requirements by extending existing technologies or replacing them with new ones.

(Alan Rae Ph D, he is associated with NanoDynamics Inc. Currently working as a Vice President of Innovations, NanoDynamics Inc. He can be reached at arae@nanodynamics.com).

Nanotechnology And Electronics

– Anil Varma

The arrival of nanotechnology has already occurred in the electronics industry with the development of elements within microprocessors that are even smaller than 100 nanometers in size. There are existing processors that are using features of 90 nanometer size. As the size gets smaller it enables quicker and more rapid processing time as well as it enables enhanced packaging of power into a given area. Nevertheless these progresses are in fact an extension of the microelectronics that is already in existence and we can safely assume that by the year 2020 the reduction of size in microelectronics is going to reach its saturation. Not only this but at exceedingly small dimensions silicon tends to electrically leak and causes short circuits.

The other side of the coin is that nanoelectronics makes available a fresh path for advancement for the electronics industry. This is in the shape of fresh materials for circuits, processors, storage of information and also means of information transfer.

(Anil Varma, Consulting Editor, The Icfai Research Centre Pune. He can be reached at avarma@iupindia.org.)

3

Samsung's Washing Machines
The 'Nano' Innovation

Roopa Umashankar

Application of revolutionary 'nano' technology in Samsung's home appliances paved way for a new genre of washing machines. Samsung's new washing machines utilized nano-sized silver particles as disinfectant to keep garments germ-free even up to month after being laundered. With the introduction of the Silver Nano washing machines, the global washing machine market has undergone a tremendous change since the inception of washing machines in the 19th century. Silver Nano washing machine utilizes a small device containing silver plates near the washtub. When electricity passes through these silver plates, it emits silver particles, which are then sprayed into the tub during the wash and rinse cycle. This machine coupled time saving along with doing away with the need of hot water to destroy germs. The new technology was also extended that the Silver Nanotechnology would be a major contributor to the expected 50% sales in the Southeast Asian region during 2004.

"Innovation based on customer feedback is key to our growth worldwide; we have taken a giant leap towards maintaining our consumers' health related safety standards through nanotechnology."[1]

– Keun Sang Yoo, Senior Manager,
Home Appliance Division, Samsung Gulf Electronics.

"Innovation through research forms the cornerstone of Samsung's philosophy worldwide and our strategy revolves around giving consumers the best in terms of both technology and quality."[2]

– B W Lee, President and CEO, Samsung Middle East and Africa.

One of Korea's leading industrial conglomerates, the Samsung Group, manages highly diversified businesses across numerous sectors. The flagship company of the Group, Samsung Electronics Co., Limited (Samsung Electronics), manufactures a variety of electronics products, prominent among them being its home appliances like refrigerators and washing machines. To satisfy the changing needs of its global customer base, Samsung Electronics has been manufacturing different types of washing machines ranging from twin-tubs to front loading [Annexure 1]. Continuous investments in R&D and prioritising user convenience has led to the development of an array of innovative products by Samsung. In 2003, the application of revolutionary 'nano' technology [Exhibit 1] in Samsung's home appliances paved way for a new genre of washing machines. Samsung's new washing machines utilized nano-sized silver particles as disinfectant to keep garments germ-free even up to a month after being laundered. With the introduction of the Silver Nano washing machines, the global washing machine market has undergone a tremendous change since the inception of washing machines in the 19th century.

Exhibit 1: Nano Definition
A nanometer is one – billionth of a meter, or roughly 75000 times smaller than the width of a human hair. A common definition of nano-technology is the technology examining and controlling materials at the molecular or atomic level.
Source: www.samsung.com

Evolution of the Global Washing Machine Industry

The earliest known washing machines were in the form of a scrub board that was invented in 1797. In 1858, Hamilton Smith of the United States patented the rotary washing machine. In 1861, the invention of wringer enabled squeezing out of excess water from wet clothing, which was later, followed by the introduction of wooden tub equipped with gears, by William Blackstone[3] in 1874. The wooden tub facilitated to stir clothes in water. In 1907, Maytag Corporation[4] introduced a wooden tub washing machine, which was followed by electric motor-driven wringer washers in 1911 that was produced by 'Upton machine Company' (It was later renamed as Whirlpool Corporation) in Michigan. By the 1930s, washing machines became automatic with the addition of motor-driven drain pumps. In the mid-1930s, a significant improvement in washing machines was witnessed with the invention of a device, which washed, rinsed and removed water in a single operation. This device was invented by a subsidiary of Bendix Aviation Corporation.[5] Setting a clock-timing device for a pre-determined length of wash-cycle was the next development in the washing machine. Another milestone was achieved in 1947 when the 'Nineteen Hundred Corporation', a predecessor of Whirlpool,[6] introduced the first top-loading automatic washing machine. By the early 1950s, a spin-dry feature was introduced in their machines by many American manufacturers to replace the wringer. General Electric in 1957 came up with a washing machine, which controlled wash temperature, rinse temperature, agitation and spin speed. A technological advancement occurred in the late 1970s with the addition of microchip to automatic washing machines, which improved the efficiency and ease of use of the appliance. In the 1980s, a number of washing machine manufacturers emerged primarily due to increasing demand as well as easy availability of manufacturing technology. Korean manufacturers like Samsung Electronics entered the electronics industry leveraging on mass production, import of foreign technology and utilizing its government support for producing electronic goods. With the rise in disposable income, as the demand for advanced washing machines increased, washing machine manufacturers started focussing on innovative features to satisfy their customers. The global consumption of washing machines also increased from the mid-1990s to the beginning of the 21st century [Exhibit 2]. Although companies from Germany, Italy and USA dominated the global washing machines market during this period, by the turn of the century, Samsung had become one of the prominent players in this market.

Exhibit 2: World – Consumption of Washing Machines, 1994-2001 (US$ 000s)

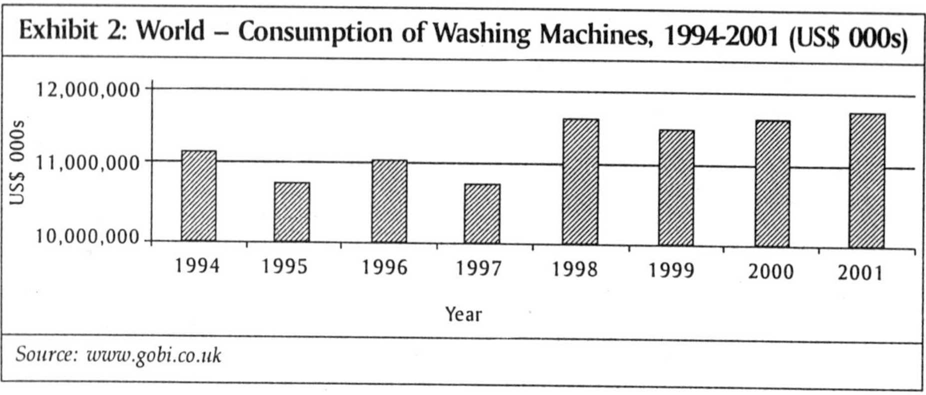

Source: www.gobi.co.uk

Samsung into the Global Washing Machine Industry

Samsung forayed into the production of washing machines in the 1970s. This venture was an attempt to diversify the portfolio of Samsung Electronics. Samsung's first washing machine (a twin-tub) was produced in 1974. In 1979, the company started exporting its twin-tub washing machines to Australia, Europe and the US. During the same year, Samsung developed the fully automatic washing machine. Samsung's lack of experience in electronics compelled it to enter into joint ventures with many suppliers of foreign technology. In addition to acquainting itself with different technologies, Samsung assembled electronics products of OEM (Original Equipment Manufacturers) buyers like Sears Roebuck, GTE, Toshiba, IBM, Hewlett-Packard. The OEM buyers in turn provided Samsung with an international market as well as design and engineering support. In 1987, a production facility was established in Suwon, Korea. Washing machine exports during the 1980s brought in substantial amount of revenues to the company, which increased from $0.7 million in 1982 to $18.1 million[7] in 1988.

In the 1990s, Samsung adopted two important strategies. One was the strategic shift from 'quantitative to qualitative growth'. The other included thrust on aggressive production. In 1991, the total production of washing machines reached 5 million.[8] Samsung developed the front-loading machine in 1993. Further, Samsung established factories in low-cost manufacturing plants like Thailand and China in 1994 and 1997 respectively to support its production facility. By 1997, the total production of washing machines by Samsung had reached 10 million.[9]

In 2002, Samsung reported a profit of $5.7 billion,[10] the major contributors being its three equally profitable divisions – consumer electronics, semiconductors and telecommunications. During the same year, Interbrand[11] and *BusinessWeek* announced that Samsung was the fastest growing brand globally and its value increased by 30% to reach $8.3 billion.[12] According to Ravinder Zutshi, Director of Samsung India Electronics – "In a category such as washing machine where technological differentiators are minimal, brand differentiators play a key role."[13] In March 2003, Samsung introduced a front-loading washing machine equipped with sterilization drying function. The machine ensured hygienic conditions by killing bacteria with the help of a high-efficiency electronic heater. It was introduced to satisfy the demand for health-promoting appliances in Seoul. Although Samsung's diversified portfolio safeguarded it considerably during the global downturn in 2003, the company's profits suffered and its turnover decreased from 9.93 trillion Won in 2002 to 9.6 trillion Won[14] in 2003. Samsung's urge to provide innovative products to its increasing health-conscious customers resulted in creating washing machines with the revolutionary "Silver Nano" technology in April 2003.

The application of 'Silver Nano' technology to Samsung's washing machines was a product of research involving 30 electronics experts and an investment of $10 million[15] in R&D. Samsung launched its 'Silver Nano' technology in its SEW-3P105A model. Silver Nano employed the sanitizing power of nano silver sized ions (a particle about 1/10,000 the thickness of human hair) to destroy airborne germs and bacteria. Commenting on the introduction of this Ag+ ('AG' stands for 'Argentum', the Latin name of Silver) washing machines, Lee Hak Soo, Senior Vice President, Samsung Electronics Digital Appliance Division, said, "Samsung has been the leader in the development of health-friendly and environment-friendly consumer electronics to address people's health concerns. Our introduction of Korea's first sterilization washing machine using nanotechnology is further testimony to the level of our technological sophistication."[16] The new washing machine utilises the sterilising effect of silver and hence when the machine is set on its 'Silver Sterilization' mode it generates billions of silver ions, which kills existing bacteria in clothes as well as those on the washing machine tub. The Silver Nano washing machine utilizes a small device containing silver plates (99.9% pure silver) near the washtub. When electricity is passed through these silver plates, it emits silver particles, which are then sprayed into the tub during the wash and rinse cycle [Exhibit 3]. Nearly 400 billion silver ions are emitted which penetrate into the fabric for sterilisation. These silver ions guard the fabric and in the process disinfect the internal part of the drum. Washing machines when treated with this groundbreaking technology act as a disinfectant and remove 99.9% bacteria present in the clothing and keep the clothes fresh for 30 days after being laundered. According to Sang Youl Eom, President and CEO of Samsung Electronics Philippines Corporation (SEPCO), "Samsung specifically created this revolutionary technology to meet the rapidly increasing consumer demand for health-related products. And Southeast Asia is the first region in the world, with the Philippines among the first countries to launch the 'Silver Nano' technology."[17]

Exhibit 3: Process of Silver Wash – Five Steps
Step 1: Electricity is passed through the electrodes to generate Silver ions.
Step 2: Tap water then flows through the electrodes carrying Ag+ (Silver) ions to the tub.
Step 3: Ag+ ionized water is then released during the wash and rinse cycle.
Step 4: Ag+ ions disinfect the clothes.
Step 5: Ag+ ions act to prevent bacteria from clinging on to the clean cloth and block their propagation into the tub.
Source: www.samsung.com

Towards the end of 2003, Samsung Electronics launched the world's biggest washing machine[18] with 10-kg capacity washing load. Commenting on the launch of the washing machine in the Middle East, Keun S Yoo, Senior Manager in the Home Appliance division, Samsung Gulf Electronics, said, "The launch of the world's biggest washing machine, equipped with the revolutionary Bigwash4D technology, is a breakthrough for the Middle East's home appliance industry."[19] Samsung also augmented its overseas production levels of washing machines in 2003. In 2004, Samsung integrated its Silver nano technology in its HA-1235A washing machine model with 10-kg capacity. This machine coupled time saving along with doing away with the need of hot water to destroy germs. Besides, the new technology was also extended to washing machines of all capacities. It was estimated that the Silver Nano technology would be a major contributor to the expected 50% sales in the SouthEast Asian region during 2004. Meanwhile, Samsung Electronics has embarked on digital convergence technology to become a major player in future in its consumer electronics division.

Sustaining and Increasing the Market Share

To drive its future successfully, Samsung Electronics intends to focus on 'market-savvy innovation'. Samsung home appliances aims to capture strategic markets like China, Europe and North America by focussing on premium products like washing machines. Subsequently, Samsung's competition with Whirlpool Corporation, General Electric Co., and Electrolux, which are already focussed on the 'premium' washing machines, is likely to increase. By 2005, Samsung intends to be a market leader in all its product segments. Commenting on the future plans of Samsung, B W Lee, President and CEO, Samsung Middle East and Africa, said, "We intend to strengthen our position by introducing a range of innovative digital products that are consistent in line with our global vision of digital convergence."[20] Samsung Electronics aims to obtain a greater domestic market share in the white goods segment, in particular for the washing machines, where competitors like LG rules the segment. The company plans to focus on creating innovative products to connect the home appliances with a digital future. Samsung Electronics introduced nano technology into the market much ahead of its major competitors like LG and Daewoo. This has given Samsung an edge over its competitors. Samsung Electronics plans to leverage its innovative technology to obtain leadership in the washing machine market. To quote Lee Hak Soo, Senior Vice President, Samsung Electronics Digital Appliance Division, "In future, we will remain in the forefront of the market by continuing to come out with new top-quality products that put the customers' health and the environment first."[21]

(Roopa Umashankar, The Icfai Business school Case Development Centre, Hyderabad. She can be reached at roopaa@icfai.org, and roopaausk@rediffmail.com.)

Endnotes

1 "Samsung launches health conscious products equipped with nano-technology", *www.ameinfo.com*, April 20th 2004.

2 "Samsung targets regional revenues of $2.7 billion in 2004", *www.samsung.com*, January 2004.

3 A manufacturer of corn planters from Indiana.

4 A commercial appliance company with its headquarters in Iowa, a state in north-central United States.

5 Founded in 1924, a Louisiana-based company.

6 A US-based producer and marketer of home appliances.

7 Kim youngsoo, "Technological Capabilities and Samsung Electronics' International Production Network in Asia", *www.brie.berkely.edu*, November 1997.

8 *www.samsung.com*

9 Ibid.

10 Ramos De Abe "Watch Out Sony", *www.cfoasia.com* , April 2003.

11 A London-based company founded in 1974 offering services like Brand Research and Brand Valuation.

12 "The Best Global Brands", *www.businessweek.com*, August 5th 2002.

13 "Samsung washing machine: Appropriating the freshness factor", *www.agencyfaqs.com*, August 18th 2003.

14 "Samsung profit dips 41 percent", *www.cnn.com*, April 18th 2003.

15 "Samsung's Silver Nano-treated Air-conditioner is literally a breath of fresh air", *www.samsung.com*, June 8th 2004.

16 "Samsung Electronics Introduces Korea's First "Silver Sterilization Washing Machine", *www.samsung.com*, April 14th 2003.

17 "Home Clean Home", *www.samsung.com*, June 8th 2004.

18 "Samsung Launches World's Biggest Washing Machine", *www.samsung.com*, September 6th 2003.

19 Ibid.

20 "Samsung targets regional revenues of $2.7 billion in 2004," *www.samsung.com*, March 2004.

21 "Samsung Electronics Introduces Korea's First "Silver Sterilization Washing Machine", op.cit.

Annexure 1: Samsung Washing Machines – Major Models

Big wash
- The Largest Capacity
- Time Saving
- Innovative High-Tech

Standard
- Saving Energy
- Quick Wash
- Eco+Wash

Slim
- New Trendy
- Sensor Cornpact
- Bio Compact

Fully Automatic
- Power Drum Inverter
- Power Drum
- UBS

Twin Tub
- New Trendy
- Classic

Source: www.samsung.com

Samsungs's Developments In Nano Innovation

– Anil Varma

Novel consumer products which make use of nanotechnology enters the market and that too at a very fast pace. The amount of consumer products using nanotechnology has developed to a very great extent. Nearly sixty percent of the product developments have been in the items of health and fitness which include products like cosmetics, lotions etc. On visiting the various sites we can see really exhaustive lists of nanotechnology products ranging from nanotech paint and diamonds, to sport items and iPhones. One of the materials in use that is most referred to is nanoscale silver. It is found in more than twenty percent of the existing products. The next most referred nanoscale material is carbon, inclusive of carbon nanotubes.

The trust of the public is the main hindrance or stumbling block in the future of nanotechnology. Governments and industries have to work to develop the confidence of the public in nanotechnology. If this is not done then nanotechnology may not find investors in the future. The growth rate of the usage of nanotechnology in consumer products and industrial application is very rapid. The products that have already been developed are just the tip of the iceberg. There is a lot more to come.

(Anil Varma, Consulting Editor, The Icfai Research Centre Pune. He can be reached at avarma@iupindia.org.)

4

Coatings: Tiny Titans

Larry Adams

Scratch resistance in high gloss coatings that are offered by these mineral oxides has reached levels that had never existed before. The scratch resistance properties of hard particles such as aluminum and silica have been well known for a long time. With nano, the scratch resistance remains and the coatings can be used in even high gloss areas. A nanocoating also has advantages over traditional resins when it comes to an ability to inhibit corrosion and act as a thermal insulator. The article also highlights that we are in the early stages of a profound industry change. Using nano-based materials could reduce the cost of applying a coating from a few dimes per article down to 1 cent per article or less.

The nanotechnology revolution which is affecting so many industries is also making a huge impact on the coatings segment, as the addition of nano-engineered particles increasingly are used to enhance coatings performance to the levels of previously unimagined. Furthermore, the nanoparticles going into coatings are reaching ever-smaller sizes, and with the reduction in size comes an increase in capabilities and opportunities to use materials in new ways.

Part of a new breed of smart coatings, nanotech coatings can make a product harder, even at extreme temperatures, more scratch and abrasion resistant, less likely to crack, and offers improved thermal, optical and electrical properties, all the while keeping its aesthetic requirements.

The secret is in the size of the particle. The smaller the particle the less likely it is that it will affect the appearance of the coatings. The smaller the particles, the better to hide scratches, to increase hardness, to repel water.

Because of these properties, designers may have some flexibility in terms of materials used – plastic perhaps, where metal was traditionally specified – because of the toughness provided by the coatings. Different kinds of materials may be used together and cured at the same time, without the need to disassemble a part to coat them separately.

From the Ground up

Nature builds from the ground up, atom-by-atom, molecule-by-molecule. The coatings industry works to replicate that feat (see the sidebar, "Making the Particle"). Metallic compounds are condensed into nanoparticles of various sizes, and then added to the coating material. At 40 nm, which is the particle size that is often used, but one that can consistently be surpassed, a slight hazing with clear coatings may occur, but the impact is generally negligible because the amount of the material is in the 0.3 percent range, suppliers say. Pigmented coatings would not have this problem. As particle size is reduced, the problem with clears is essentially eliminated.

As the particles continues to be broken down in size, they become much more highly active surface-wise, says Robert McMullin, manager of nanotechnologies, BYK Chemie, Wallingford, Conn. He adds that a particle that is 2 microns in size is going to have a small surface area, but if that 2-micron particle is broken down to 3 million or 4 million, 25 nm particles, then there are a huge number of little particles that are impregnated in the coating from surface to substrate.

Scientists at BYK Chemie Work with Coating Materials

Coatings from Industrial Nanotech are Applied to a Flat Surface

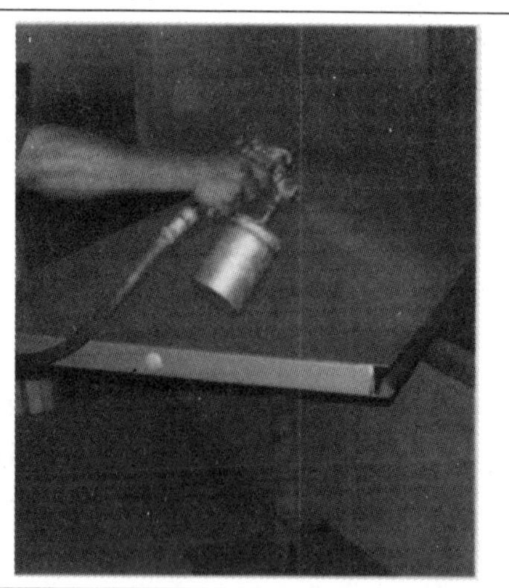

A Scanning Electron Microscope at Sandia National Laboratory Captured this Image of Industrial-Nanotech's Nansulate Coating

Top Surface Higher Magnification

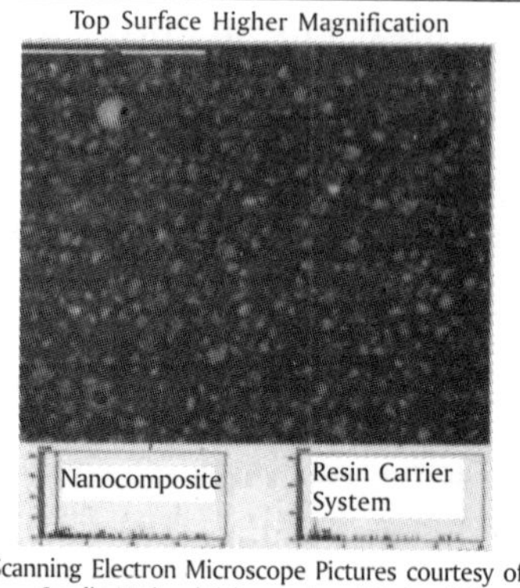

Scanning Electron Microscope Pictures courtesy of Sandia National Laboratories of Industrial Nanotech, Inc. Nansulate ™ coating

"This is how they become so active, you have all of these little particles spread uniformly through the coating," he says. "If all particles are on the surface, once you get past that level you are going to get micro-degradation and as soon as you eliminate that top coating than you are back to nothing. Take dispersion. The nanoparticles are dispersed throughout the coating so that they can provide scratch protection from substrate to the top layer. If the nano material is only on the surface, microdegradation would soon make the piece susceptible to damage."

A Pigmented Coating from Ecology Coatings is Used on this Cylinder, Nanoparticies do not Alter the Look of Pigmented Coatings

Coatings Get Tough

For scratch and abrasion resistance, alumina (Al_2O_3) and silica (SiO_2) particles have been found useful for this application, says McMullin. BYK has teamed up with Nanophase Technologies, Romeoville, Ill., a developer of nanoparticies.

He says that the scratch resistance in high-gloss coatings that are offered by these mineral oxides has reached levels that had never existed before. To illustrate this, dry scrub abrasion tests using steel wool or aluminum-oxide paper ranging from 3 microns to 9 microns running up to 2,000 cycles were conducted. Gloss readings of a surface that did not have these coatings were about a third less than materials that did have the nanotech coatings. (See Figure 1).

The scratch resistance properties of hard particles such as aluminum and silica have been well known for a long time, says McMuliin. However, when large particles, those that are several microns in diameter are used, the coatings experience a loss of gloss and transparency. For instance, titanium oxide tends to whiten a coating. Because of that, these non-nano scratch materials are best used in pigmented, opaque or flat, no-gloss formulations.

Figure 1: Abrasion Testing by BYK Chemie Reveals that Gloss Levels will Remain Approximately the Same when Using a Nanotech Coating

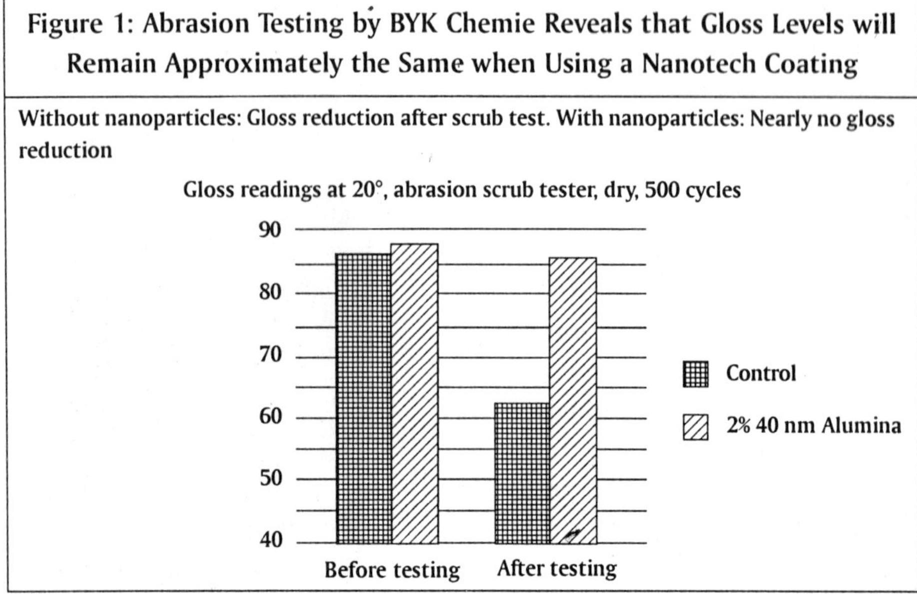

With nano, the scratch resistance remains and the coatings can be used in even high gloss areas. In terms of hardness, diamond is the hardest on a ratio index at 10, and aluminum comes in at 9, so coatings with this material make up is fairly abrasion resistant. Silica, another material often used to protect against scratches, rates 7, so it has a little less resistance. On the other hand, aluminum has a refractive index of 1.7, which is higher than a normal resin and silica has a refractive index of 1.6, dead center in the middle of most refractive levels for resins. So, aluminum delivers hardness, but silica has more scratch-hiding properties.

By combining the two materiais, the best of both coatings can be achieved. (See Figure 2.)

This scratch and abrasion fighting ability may offer designers new material choices. Plastics are not as durable as most metals. If a designer wants to use more plastic parts, then better surface protection is required.

"A coating on top of a piece of plastic that has abrasion resistance is going to be able to function closely to a steel shell or a steel insert," says McMullin. "It gives designers the capability of utilizing materials other than metal and make

Figure 2: Tests by BYK Chemie Shows that Scratch Resistance is Improved when Alumina Nanoparticles and Silicon Surface Additives are Combined

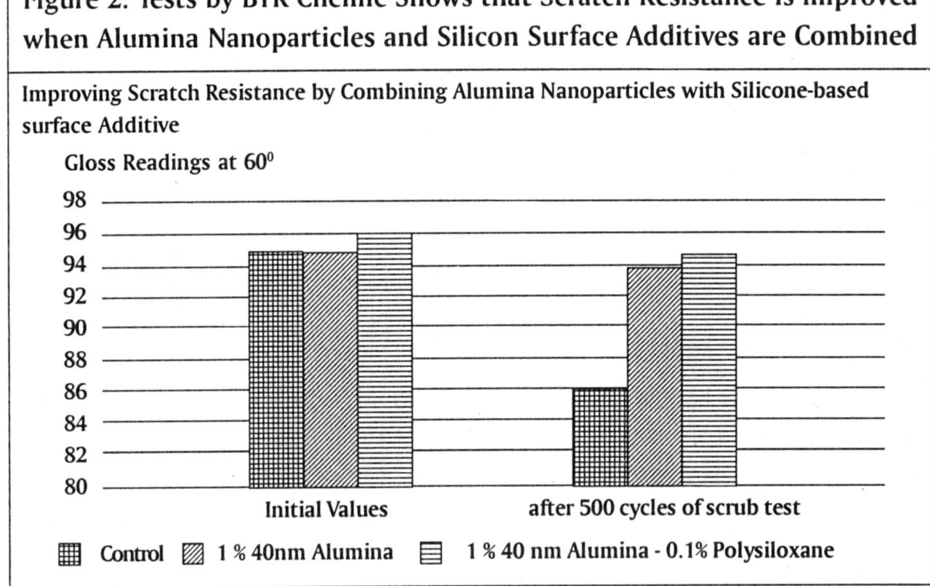

them just as functional. Wouldn't it be interesting to see a plastic shell of a refrigerator, rather than a metallic shell?"

Fighting Corrosion

A nanocoating also have advantages over traditional resins when it comes to an ability to inhibit corrosion and act a thermal insulator.

Nanoparticles are hydrophobic in nature, says, Francesca Crolley, vice president of operations and marketing for Industrial Nanotech, Naples, Fla., which means that they can repel water and inhibit corrosion.

For instance, Industrial Nanotech's Nansulate coatings were subjected to automotive corrosion testing earlier this year. The coating, which was tested under the GM 9540 P automotive corrosion test standard, was subjected to, and passed, more than 16 cycles of acid splashes and salt sprays. "Typically," says Crolley, "8 cycles is considered passing."

Industrial Nanotech coatings such as Nanuslate are often used in high-temperature environments. LS Industries, Wichita, Kan., uses Nansulate on industrial ovens to reduce surface temperature to worker-safe levels. These ovens are used to clean engines by burning off residue at 800DegF. The external temperature of the oven after a 7-mil coating dropped from I68DegF to between lOODegF and 115DegF, which was considered safe. Additional material would have reduced the external temperature even lower.

The material also works to prevent corrosion under insulation that is caused by the thermal differential between the coating and the equipment. The nanocomposite that constitutes approximately 70 percent of the dry-film thickness of Nansulate is hydrophobic, which acts to repel moisture build-up.

Light Curing

UV curing systems for coatings are increasingly used in finishing systems today because they speed up the process and eliminate production bottlenecks. Of course, such systems require the availability of a UV-curable coating. Fortunately, a number of companies now offer UV coatings. BYK-Chemie, for instance, offers the NANOBYK-3600, 3601, and 3602 that are based on alumina particles with a mean particle size of 40 nm and can be UV cured.

One of the more acclaimed coatings to be developed in recent years comes from Ecology Coatings, Akron, Ohio. This company produces a nanotech coating that is 100 percent solid and has been dubbed "liquid solid." It goes on as a liquid and UV cures within seconds. Overspray is completely reclaimable. "Everything that is in the liquid becomes solid," says Sally Ramsey, chief chemist and company founder, "You don't have water or solvent carriers. What you spray is what you get. What goes on wet, ends up dry."

While UV curing is generally a fast process, this coating's proprietary nanotech structure makes it even quicker to cure, which she says Is the product's biggest advantage. She says that the coatings' structure requires 75 percent less energy and 99 percent less time to cure.

It is important to know, however, where and when these types of coatings are applicable.

For example, many parts today are designed with complex geometries and hidden surfaces. UV cure coatings are cured by line-of-sight transmission of light; parts or assemblages with complex geometries are not suggested for these types of coatings.

On the other hand, assemblages or composite parts that contain multiple materials, including temperature-sensitive plastics, might be good for UV curing. For example, some assemblages must be disassembled, coated and then reassembled because of temperature sensitivity issue with one or more of its components, says Ramsey. "Because we raise the temperature only a few degrees, you don't have to disassemble. If you have a metal part with a rubber filler, you don't necessarily have to disassemble, because you are not going to get enough heat to ruin the rubber."

The nanotech coatings also require less material, she says. "If using one of our coatings that was non-nano, you'd probably want to put on at least 1.5 mils for good coverage with a pigment," says Ramsey. "But, if you were using the nano version, you would only need to put on 1.2 millions."

In the course of benchmarking, DuPont tested Ecology Coatings' materials and was impressed enough to enter into a licensing agreement. Ecology's promotional materials include a testimonial from Robert Matheson, DuPont's technical manager for strategic technology: "We are in the early stages of a profound industry change. Using Ecology Coatings' nano-based materials could reduce the cost of applying a coating from a few dimes per article down to 1 cent per article or less."

If Matheson's experience is the norm, cost of application will continue to fall. Shrinking, too, is the size of the particles. However, smaller particles can bring higher prices so, as with many other specification decisions, the benefits of higher performance must be weighed against higher costs.

But, there is one thing for certain, the creation of smaller and smaller nano-sized particles will continue unabated. Where once the micron was infinitesimal, now it is not even a boulder, but more of a mountain. "Today, we are starting to think of 40 nm as our boulders," says McMullin. "We are starting to break the nanometers down consistently into 25 nanometers, those are becoming very adequate for commercialization."

(Larry Adams is a Managing Editor at appliance design and can be reached at adamsl@bnpmedia.com).

Making the Particle

Nanoparticles can be made in a variety of methods, but many rely on thermally breaking down metal compounds, adding vapor and collecting the particles that condense as the mixture cools.

Nanuptiase Technologies Corp, Romeoville, Ill, uses two methods to make its nano particles. These are:

Physical Vapor Synthesis

Ttie first nanopowder manufacturing process developed and scaled-up at Nanophase Technologies Corp, was the patented physical vapor synthesis process. In this process, arc energy is first applied to a solid precursor (typically metal) in order to generate a vapor at high temperature. A reactant gas is then added to the vapor, which is then cooled at a controlled rate and condenses form nanoparticles. The nano-materials produced by the PVS process consist of discrete, fully-dense particles of defined crystallinity.

This method typically produces particlss with average sizes ranging from 8 nm to 75 nm. Nanophase Technologies use the PVS process in the commercial-scale production of NanoGard Zinc Oxide and NanoTek Aluminum Oxide. In addition, this process has been used to generate additional materials such as a variety of doped zinc oxides, selected rare earth and transition metal oxides, and transparent conductive oxides such as antimony-tin oxide and indium-tin oxide,

NanoArc Synthesis

The second-generation nanopowde manufacturing process developed at Nanophase Technologies is the patent-pending nanoarc synthesis process. Like the PVS process, the NAS process uses arc energy to produce nanoparticles. The NAS process, however, is capable of using a wide variety precursor formats and chemical compositions, thereby greatly expanding the number of materials that can be manufactured as nanopowders at commercial scale.

The nanomaterials produced by the NAS process consist of discrete, fully-dense particles of defined crystallinity. This method has been used to produce particles with average sizes ranging from 7 nm to 45 nm.

An enhanced capability of the NAS process is its ability to process complex multi-component materials. This process has demonstrated the ability to produce homogeneous mixed metal oxide nanopowders where the component materials form solid solutions with well-defined single crystalline phases. Nanocrystalline metal oxides having up to four metallic elements have heen successfully produced.

The NAS proess has the capability to produce a wide variety of single-phase pure and mixed rare earth oxides, as well as pure and mixed transition metal oxides and main group metal oxides. The materials produced by this process have application in ultrafine polishing and chemical-mechanical planarization, catalysis, fuel cells, electronic materials, and advanced imaging.

A unique capability of the NAS process is on site particle surface modification. This capability has heen used to produce materials, at commercial volumes, which can be processed to yield extremely stable dispersions, such as Cerium Oxide SG, which has found particular utility in ultra fine ploishing and chemical-mechanical planarization applications.

Source: Nanophase technologies Corp.

Section II

Specialized Applications

5

BIPHOR – Nanotechnology
For Waterborne Paint Improvement

Fernando Galembeck, Maria do Carmo V M da Silva, Renato Rosseto, Gilmar O Pinheiro and João de Brito

The paint industry has come up with a new product called BIPHOR the basis of which is the traditional concepts of nanotechnology. It is an exceptional amalgamation of the advantageous functions of waterborne paint which has been described and explained in this article. The article also highlights that BIPHOR in waterborne paints brings about enhanced visual performance and durability at a reduced cost.

BIPHOR™ is a new product for the paint industry based on contemporary nanotechnology concepts that are now reaching the semi-industrial production stage. It performs an outstanding combination of desirable functions in a waterborne paint that are described and explained in this paper. The scientific basis for the use of BIPHOR was established during work performed at Unicamp,[1-7] followed by

This Article earlier appeared in Paint & Coatings Industry Magazine, a BNP Media Publication, www.pcimag.com

joint process and product development made by Bunge Fertilizantes S.A. and Unicamp since the mid-90s.

What is BIPHOR?

BIPHOR is a new family of aluminum phosphates or polyphosphates made by a patented wet chemistry process.[8] It is a "green chemistry," zero-effluent product made under mild temperature and pressure conditions that do not create any environmental problems during the fabrication process.

Due to its chemical nature, BIPHOR residues in the paint industry or in the final user location may be sadly discarded in the environment as a fertilizer component. It is produced as slurry as well as a dry powder. In both cases it is easily dispersed in water, forming stable dispersions that have stable theological properties.

BIPHOR is completely different from existing aluminum phosphates or polyphosphates in many aspects.

- It does not have a definite stoichiometry. thus it is better thought of as a crosslinked ionic polymer than as a salt.
- Since the stoichiometry is not definite, various grades of BIPHOR can be prepared by changing the fabrication process, and thus the final product composition.
- BIPHOR is made under very rigid pH control, and it is nearly neutral, thus avoiding environmental and toxicological problems.
- It is also exempt from corrosion problems associated with some aluminum phosphates found in the marketplace and used in the transformation of iron oxides into iron phosphate in metal-protecting paints.
- The non-stoichiometry, together with the non-crystallinity of BIPHOR (both slurry and powder) and the carefully controlled water content of dry

BIPHOR powder, allow for easy swelling control that is essential for its performance, following the mechanism that will be described.

- The nano-sized particles are easily dispersed and are stable towards settling, which is very important for making uniform paint dispersions with BIPHOR.
- BIPHOR nanoparticles are strongly compatible with latex particles, by the mechanisms of capillary adhesion (in the dispersion drying stage) followed by ion-cluster mediated electrostatic adhesion (in the dry film). This compatibility is so high that bicontinuous networks are formed in many cases.
- BIPHOR is also strongly compatible with many other particulate solids commonly used as paint fillers, such as the various silicates, carbonates and oxides found in formulated waterborne dispersions. This contributes to the cohesion and strength of the dry paint film.

Typical chemical composition data of the current product are shown in Table 1. Together with diffraction data (Figure 1), the infrared spectrum (Figure 2), electron and atomic force microscopy analysis (Figures 3 and 4) and thermogravimetric data (Figure 5) these results show that BIPHOR is a hydrous, non-crystalline and neutral aluminum phosphate made out of nano-sized particles. Particle size (in the slurry) is in the 200-2,000 nm range, as determined by dynamic light scattering.

Table 1: Elemental Ratios of some BIPHOR Grades as Determined by X-ray Fluorescence Using Futiddmental Pararrieiers

Grade	P	Al	S	Si	Fe	Ca
1	1	0.800	nil	0.067	0.0006	0.0005
2	1	0.820	nil	0.049	0.0005	0.0014
3	1	0.769	0.026	0.058	0.0007	0.0012
4	1	1.26	0.54	0.04	0.019	nil

Figure 1: X-Ray Diffractogram of a BIPHOR Grade. The Broad Bands Show that this Powder is Non-Crystalline or Else, it is Amorphous

Intensity (CPS)
600
500
400
300
200
100
0
0 10 20 30 40 50 60 70 80
20

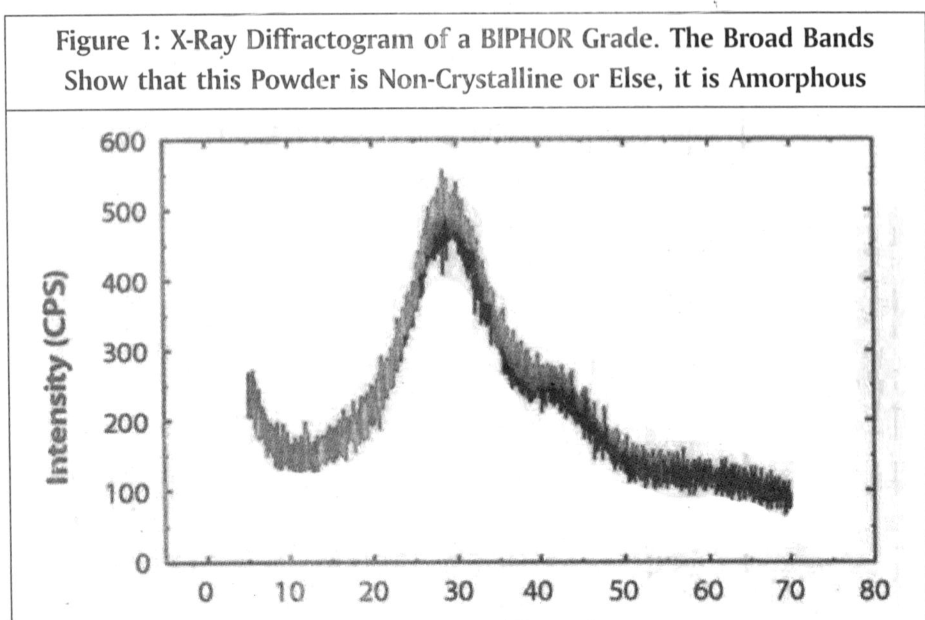

Figure 2: Infrared Spectrum of a BIPHOR Grade in KBr. The Large Band at 3700-2700cm^{-1} is due to the Extensive Hydration of BIPHOR Particles

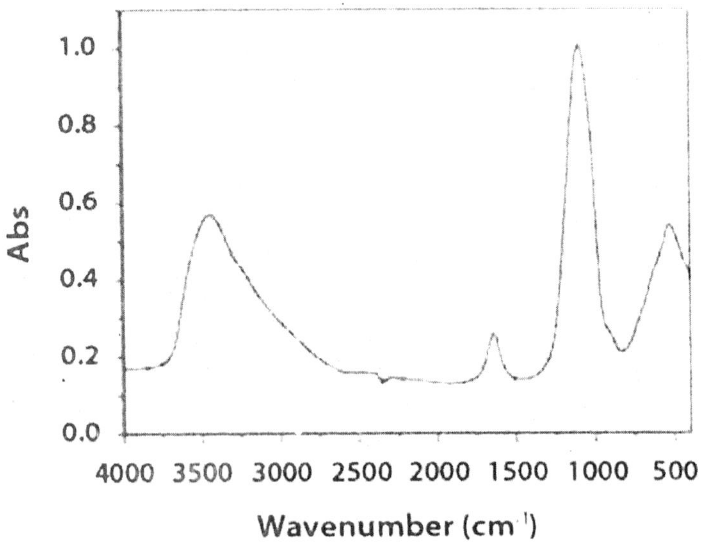

Abs
1.0
0.8
0.6
0.4
0.2
0.0
4000 3500 3000 2500 2000 1500 1000 500
Wavenumber (cm⁻¹)

Figure 3: Electron Micrograph BIPHOR Sample Showing the Nano-sized Particles

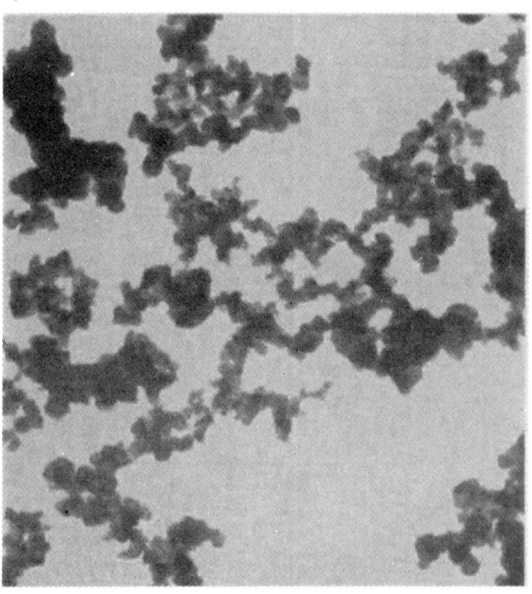

Figure 4: Atomic Force Micrograph of a BIPHOR Powder Showing the Packed Nano-sized Particles

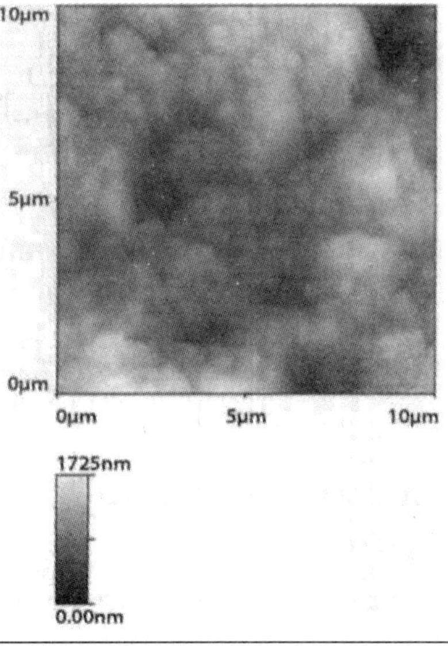

Figure 5: Thermogram of a BIPHOR Grade. The Weight loss is due to the Evaporation of Water

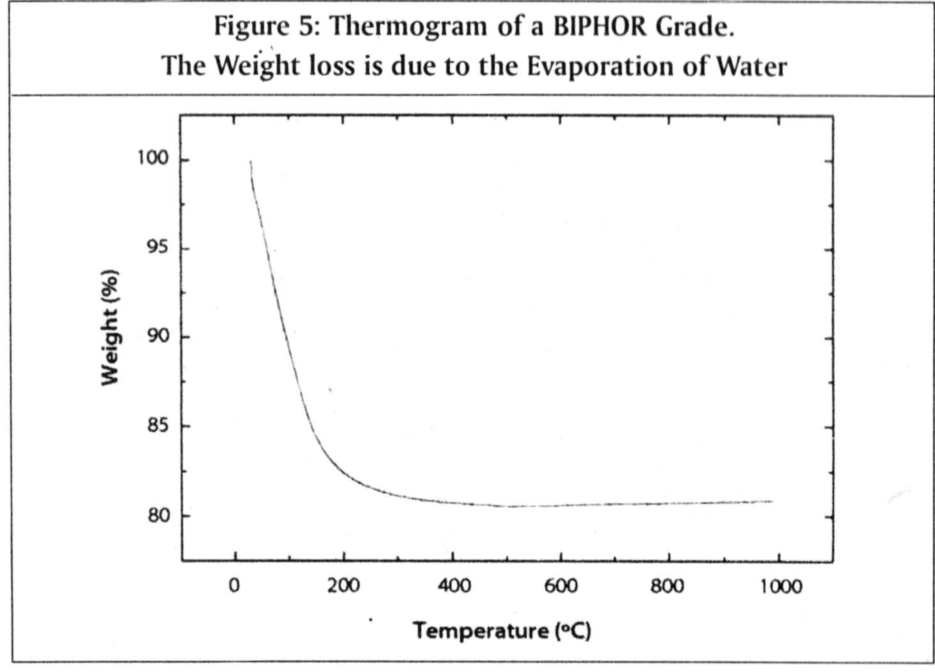

How BIPHOR Works

To understand how BIPHOR performs within a drying paint film, it is interesting to compare it to the current industry-standard white pigment, titanium dioxide. A basic TiO_2 waterborne paint is made out of a suitable latex dispersion and pigment particles. The latex particles are responsible for making a coalesced film filled with the pigmented particles, which are responsible for the film hiding power. Many additives are also used: Inorganic fillers, which decrease the requirements of resin and pigment; coalescing agents that improve resin film formation; and dispersants and rheological modifiers that prevent pigment and filler caking, thus improving the paint shelf-life together with the rheological paint properties.

A typical dry paint film can thus be represented as shown in Figure 6. The pigment and filler particles are dispersed in the resin film. In this case, hiding power is largely dependent on the particle refractive indices and sizes. TiO_2 is currently the standard white pigment because of its high refractive index and because of the absence of light absorption in the visible region.

Figure 6: Schematic Representation of a Tio_2-pigmented Resin Film. Pigment and filler particles are scattered throughout the film and they backscatter incident light due to the large refractive index difference between the resin and the particles that have an optimized diameter

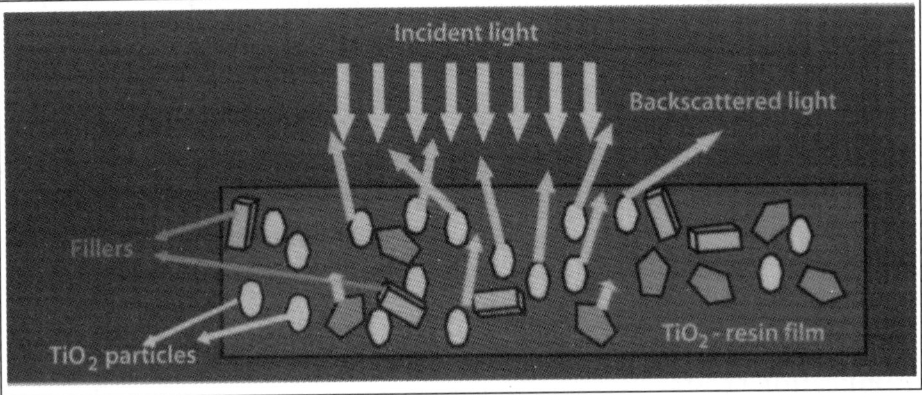

A typical dry film of a paint formulated with BIPHOR is represented in Figure 7. There are many large differences compared to the previous case.

Figure 7: Schematic Representation of a BIPHOR Resin Film with Fillers. Pigment particles as well as closed pores are scattered throughout the film, and they backscatter incident light due to the large refractive index difference between the resin and the particles or closed pores.

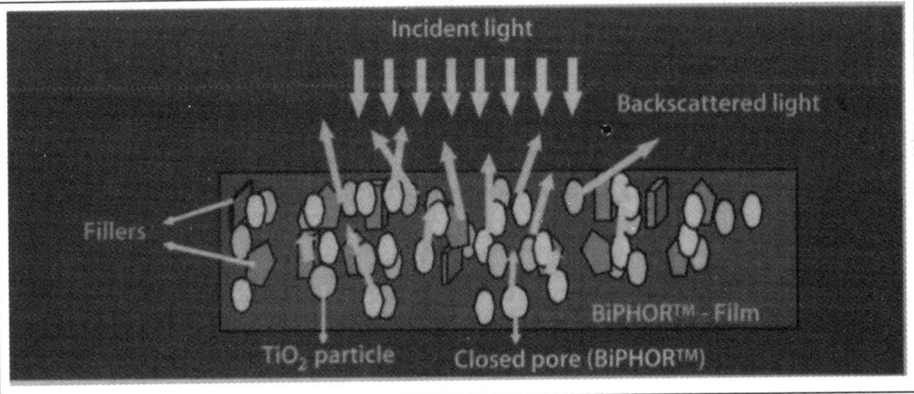

- The film is not just a resin film; it is formed by enmeshed resin and aluminum phosphate. It is thus a nanocomposite film that combines two interpenetrating phases with different properties to achieve synergistic benefits concerning film mechanical properties and resistance to water and to other aggressive agents. This is a new concept for the paint industry but it is already largely used in the development of new, high-performance materials generally described as nanohybrids or nanocomposites.[(9)]
- Good film hiding power is obtained at lower TiO_2 content because the film contains a large amount of closed pores that also scatter light. Moreover, if a TiO_2 particle is adjacent to one of these voids, it will scatter much more than if it is fully surrounded by resin, due to the larger refractive index gradient. This creates a synergism between BIPHOR and TiO_2 as far as the hiding power is concerned.

Application Tests

In these tests, a standard market formulation of a semi-matte acrylic paint was chosen, and TiO_2 was progressively replaced with BIPHOR. Water content and other paint components were adjusted as required. The majority of the modifications in the formula are related to a decreased use of thickener/rheology modifier, dispersant, acrylic resin and coalescing agent.

Table 2 describes an example of one of the standard formulas used in this work, together with the corresponding BIPHOR formula. The standard formula in Table 2 is currently being used by one of the market leaders in the Brazilian architectural paint market.

Table 2: A Standard Formula Currently Used in the Brazilian Market, and the Corresponding Formula Using BIPHOR. The Amounts are Given in Grams

	Standard Formula	Formula prepared using BIPHOR slurry
Water	839.79	361.86
Propylene glycol	30.00	30.00
Thickener/rheology modifier	84.00	4.50

Contd...

Contd...		
Antifoaming agent	0.60	1.17
Sodium tetra pyrophosphate	0.87	9.00
Anti-oxidant	0.87	0.90
Dispersant	20.94	11.00
Amine	-	5.00
AFT anionic	7.86	7.86
Bactericide	4.50	4.50
Fungicide	4.50	4.50
Ammonium hydroxide 25%	7.11	15.00
Titanium dioxide	534.00	267.00
Kaolin #325	169.50	169.50
$CaCo_3$ nat. micronized	161.28	161.28
Dolomite #325	300.00	300.00
Aluminum silicate #1000	60.18	60.18
BIPHOR slurry 35%	-	763.00
Acrylic resin	735.00	591.00
Antifoaming/mineral spirit	9.00	6.00
Coalescing agent	60.00	43.47
Total (grams)	**3030.00**	**2816.72**

In the formula in Table 2 a replacement of 50% TiO_2 (on a weight basis) was made, obviously keeping the opacity and whiteness conditions of the dry film. In addition, the other properties of BIPHOR as a rheological modifier and also as a film-structuring agent were explored.

Comparison between the two formulas shows that the basic intrinsic properties of BIPHOR will lead to additional cost reduction beyond that derived from the replacement of TiO_2 pigment only. Moreover, these gains are obtained while producing excellent final performance in the applied paint film. Table 3 and 4 show the evaluation results of the two architectural semi-matte acrylic paints referred to in Table 2, conducted by two independent laboratories. The formulation made by replacing 50% by weight of TiO_2 with BIPHOR is comparable to the standard paint in gloss, reflectance, opacity, whiteness and yellowness. The appearance of dry drawn films formulated with BIPHOR (50%) and TiO_2 (100%) is illustrated in Figure 8.

Table 3: Properties of Bunge Paints[a]		
	Standard Formula	**Formula prepared using BIPHOR slurry**
Description	100% Ti[O.sub.2]	50% BIPHOR Slurry
Non-Volatile (%)	52.6	50.7
Stormer Viscosity	100	113
0.003" Drawdowns		
85^0 Gloss	1.2	0.90
Contrast Ratio	0.8972	0.9022
Reflectance (white)	0.8887	0.8914
Yellowing Index (D1925)	4.58	4.59
Yellowing Index (E313)	3.97	3.99
Yellowing Index (E313)	75.13	74.88
0.0015" Drawdowns		
Contrast Ratio	0.745	0.7279
Reflectance	0.8682	0.8676

[a] Source: Stonebridge Technical Services, 6223 Linden Road, Fenton, MI.

Table 4: Properties of Bunge Paints[a]			
Slurry		**Standard Formula**	**Formula prepared using BIPHOR**
Description		100% Tio_2	50% BIPHOR Slurry
Hiding	Percent		
At 10 m^2/L		94.1	94.2
At 6.6 m^2/L		97.6	97.2
At 6.6 m^2/L	Percent		
Reflectance		91.2	91.3
Whiteness Index		80.9	81.2
Yellowness Index		3.8	3.8
Gloss-60^0	Units	4	2
Sheen-85^0		2	2
Scrub resistance	Cycles	280	260

[a] Source: DL Labs, Inc., 74 Kent Street, Brooklyn, NY.

Figure 8: Visual Comparison of Cards Painted with Standard Paint and BIPHOR-based paint, Control; 100% TiO_2, 50% BIPHOR

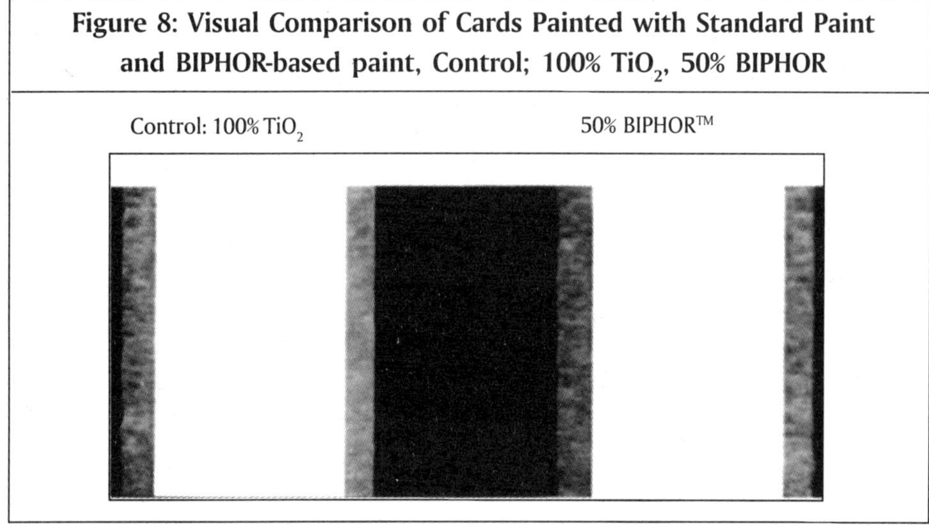

Several field test applications were also made, and an example is shown in Figure 9. Some of these application tests have been monitored for quite some time now and they are revealing great color stability and resistance to chalking. This can be attributed to the complete photochemical inertness of BIPHOR (as compared to TiO_2, bringing in a bonus for those paints formulated with BIPHOR.

Figure 9: A Field Application Test – A House Painted in the year 2004 with 3 BIPHOR-based Paint in Cajati. This City is Surrounded by Remnants of the Mata Atlântica Rain Forest in one of the Warmer, Sunnier and Most Humid Areas in Brazil

Conclusion

BIPHOR is now completing the development stage with outstanding results. The introduction of BIPHOR in waterborne paints leads to improved visual performance and durability at a reduced cost.

Acknowledgments

FG acknowledges PADCT/CNPq for supporting the Instituto do Milênio de Materiais Complexos (Millenium Institute for Complex Materials).

(Fernando Galembeck, Maria do Carmo V M da Silva, Renato Rosseto, are associated with Rosseto lnstituto de Quimica – Universidade Estadual de Campinas, Gilmar O Pinheiro and João de Brito with IIILJ where as JoaodeBrito is associated with rcrtiiizanles S,A,,SdO Paulo).

References

1. Beppu, M.M; Lima, E.C.D.; Sassaki, R.M.; Galembeck, F. Self-opacifying aluminum phosphate particles for paint film pigmentation.; *J. Ctgs. Technol.* 1997, 69 (867): 81-88.
2. Lima, E.C.D.; Beppu, M.M.; Galembeck, E; Valente, J.F.; Soares, D.M. Non-crystalline aluminum polyphosphates: Preparation and properties. *J. Brazilian Chem. Soc.* 1996, 7 (3): 209-215.
3. Beppu, M.M.; Lima, E.C.D.; Galembeck, F. Aluminum phosphate particles containing closed pores. Preparation, characterization, and use as a white pigment. *J. Colloid Interface Sci.* 1996, 178 (1): 93-103.
4. Monteiro, V.A.D.: de Souza, E.E : de Azevedo, M.M.M.; Galembeck, F. Aluminum polyphosphate nanoparticles: Preparation, particle size determination and microchemistry. I. Colloid Interface Sci. 1999, 217 (2): 237-248.
5. De Souza, E.F.; da Silva, M.D.C.V.M.; Galembeck, F. Improved latex film-glass adhesion under wet environments by using an aluminum polyphosphate filler. *J. Adhes. Sci. Technol.* 1999.13 (3): 357-378.
6. De Souza, E.F.: Bezerra, C.C.; Galembeck, F. Bicontinuous networks made of polyphosphates and of thermoplastic polymers. Polymer 1997, 38 (26): 6285-6293.
7. Lima, E.C.D.; Beppu, M.M.; Galembeck, F. Nanosized particles of aluminum polyphosphate. Langmuir 1996, 12 (7): 1701-1703.
8. Galembeck, F. "Produto e processo de fabricao de um pigmento branco baseado na sintese de particulas ocas de ortofosfato ou polifosfato de aluminio", Brasil, PI0403713-8, 2004.
9. Giannelis, E.P. Polymer layered nanocomposites. Adv. Mater. 1996, 8 (1): 29-35.

6

Nanocomposites

Jan H Schut

Skeptics who may doubt that nanocomposites have yet proven themselves commercially viable materials ranging from auto parts and precision moldings to wire and cable jacketing, there is plenty of evidence that 'nanos' are the beginning to live up to their promise. These applications demonstrate that just a pinch of these tiny particles can cut weight and cost compared with higher loadings of conventional fillers. The benefits include improved mechanical properties, scratch resistance, barrier properties, fire resistance and dimensional stability, plus faster extrusion and molding. The author explains that early applications have been motivated by cost savings relative to conventional glass and talc-filled resins. Nanomaterials weigh less, mold more easily on smaller presses, process dramatically faster and produce better properties. The most interesting potential application for nanoclays is thin-walling for lightweight parts. So far, people have used nanoclays to save money, but they haven't taken full advantage of nanotechnology to reduce wall thickness. Since nanocomposites are stiffer but lighter than glass or mineral-filled plastics and also enhance flow, one can reduce part weight and thickness.

Source: www.ptonline.com © Gardner Publications, Inc. Reprinted with permission from Plastics Technology, February. 2006.

This article highlights that they're out of the lab and into the real world. In applications from automotive to agriculture, materials handling, and electrical cable, nanoclay blends are lowering costs and raising performance. The market has already recognised how nano-materials enhance performance.

For skeptics who may doubt that nanocomposites have yet proven themselves commercially viable materials, last year's Nanocomposites 2005 Conference in San Francisco presented plenty of evidence that "nanos" are beginning to live up to their promise. Organized by Executive Conference Management, the conference included reports on a bevy of new commercial applications in Europe and North America, ranging from auto parts and precision moldings to wire and cable jacketing.

These applications demonstrate that just a pinch of these tiny particles can cut weight and cost compared with higher loadings of conventional fillers. The benefits include improved mechanical properties, scratch resistance, barrier properties, fire resistance, and dimensional stability, plus faster extrusion and molding.

News from the conference included new alloys that disperse nanoclays more efficiently, new synthetic nano-additives, and a new naturally occurring nanotube. There was also word of new twin-screw lab extruder that makes test batches as tiny as 3 cc and an infrared test method to verify nanoclay exfoliation and orientation.

Cost and Performance

Early applications have been motivated by cost savings relative to conventional glass- and talc-filled resins. Nanomaterials weigh less, mold more easily on smaller presses, process dramatically faster, and produce better properties. For example, Nanoblend MB (Melt Blend) concentrate, a nanoclay masterbatch introduced in 2003 by PolyOne Corp., is used commercially in a TPO automotive rocker panel, replacing talc-filled TPO. Nanoblend has better low-temperature ductility and lower specific gravity. Because of the lower amount of nano-filler, color matching was less expensive with Nanoblend. That plus the lower density, led to a projected savings of over $42,000/yr, despite somewhat higher material cost/lb.

PolyOne's Nanoblend LST (Light-Stiff-Tough) compound has been used commercially since 2004 in a material-handling tray that requires tight dimensions for use on an automated assembly line. The PP-based Nanoblend tray replaces a PPO alloy and is reportedly flatter, has lower coefficient of thermal expansion, and saved the processor $24,900 on a short production run because it trimmed cycle time from 90 sec to 60.

Arthur Fritts, president of NanoSperse LLC, a two-year-old compounder devoted exclusively to nano-materials, cautions against considering nano-materials just for cost savings. "Long-term growth in nanocomposites won't come from substituting a cheaper material," he says. "That has actually slowed nano developments. Growth will come from enhancing value."

In fact, the market is already recognizing how nano-materials enhance performance. An irrigation emitter for landscape and agricultural pipe, developed by Geoflow in San Francisco and molded by the Toro Co., in Bloomington, Minn., uses the nanoclay platelets in PolyOne's Nanoblend MB concentrate to ensure slow release of herbicide from the plastic into the soil. Nanoblend is dry mixed with pellets of Geoflow's herbicide, then added to LLDPE at the hopper of an injection press. That part costs 20% more than a part without nanoclay, but it extends the product guarantee from 10 years to 20.

Nanoclay in this Geoflow LLDPE drip emitter for irrigation tubing ensures timed release of herbicide from the plastic. PolyOne's Nanoblend MB nanoclay concentrate adds 20% to the cost but doubles product life

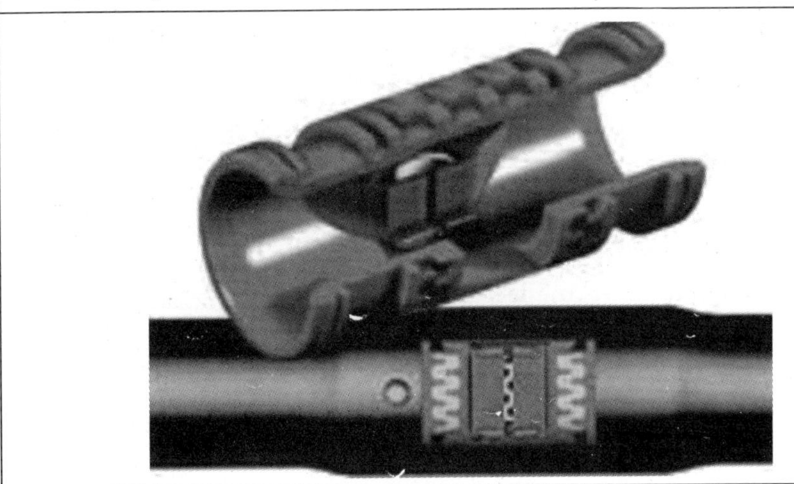

"The most interesting potential application for nanoclays is thin-walling for lightweight parts," in the view of Suresh Shah, senior technical fellow at Delphi Corp., Troy, Mich. "So far, people have used nanoclays to save money, but they haven't taken full advantage of nanotechnology to reduce wall thickness. Since nanocomposites are stiffer but lighter than glass- or mineral-filled plastics and also enhance flow, you can reduce part weight and thickness."

One way in which nanoclays improve material properties is through compatibilizing resin alloys. Putsch in Germany partnered with nanoclay supplier Sud-Chemie AG to develop a new alloy of normally incompatible PP and PS. Called Elan XP, it uses 1% to 2% of Sud-Chemie's Nanofil to compatibilize 60-80% PP with 20-40% PS. Elan XP was commercialized three years ago for an interior air vent for the Audi A4 and a Volkswagen van. The unpainted part replaces painted ABS. The hard PS component provides scratch resistance and a luxurious surface feel.

Nanocomposites have scored early Gains in automotive. A heater vent for the Audi A3 gains scratch resistance from Putsch GmbH's Elan XP compound of PP and PS, compatibilized by a Sud-Chemie nanoclay. It replaces painted ABS. Micrograph below shows finely dispersed PS domains in PP matrix

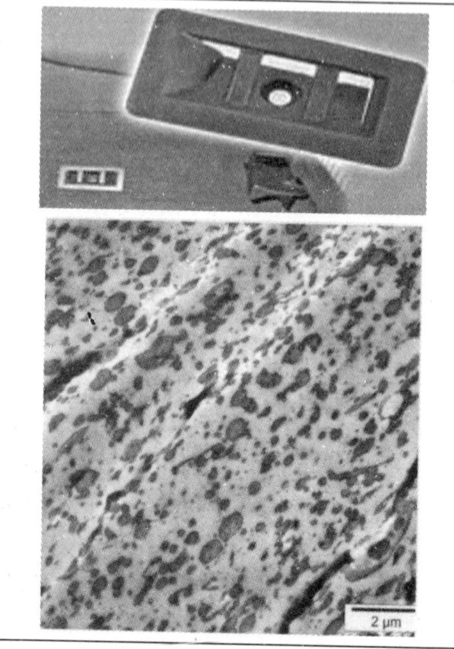

Nanoclay also has proved effective as a flame retardant: It improves UL ratings by eliminating dripping and promoting formation of stable char. For example, Sud-Chemie's new Nanofil SE 3000 nanoclay is used commercially in an EVA/PE compound to provide halogen-free flame retardency in wire and cable coatings. At only 3% to 5% loading, it improves flame retardency enough that ATH or magnesium hydroxide flame retardants can be cut from 65% to about 52% by weight. Mechanical and surface properties improve, and extrusion rates are higher, Sud-Chemie reports.

Nano-materials have been used about two years by three large European cable companies: Kabelwerk Eupen in Belgium, Nexans in France, and Draka Cable in the Netherlands. Nexans' Berk-Tek unit in the U.S., with offices in New Holland, Pa., is believed to have the first nanocomposite cable jacketing in North America. Nexan introduced it 18 months ago for its plenum cable used in office buildings.

Still another function of nanoclay is as a barrier enhancer. Early attention focused on food and beverage packaging, but automotive fuel systems are an emerging application. Ube America has developed a coex barrier fuel line with 2% nanoclay in nylon 6. Its gasoline barrier is reportedly five times greater than neat nylon.

In-Situ Compounding

PolyOne commercialized a new family of nylon Nanoblends made by in-situ polymerization of caprolactam with nanoclay. PolyOne calls them "alloys" because the caprolactam is chemically bound to the nanoclay before and after polymerization. This prevents the possibility of reagglomeration of clay particles in subsequent processing. Reagglomeration effectively negates the potential benefits of a nanocomposite.

Nano Materials are Moving into Applications like Nexans' Plenum Cable Jacketing, the First Such Product in the US

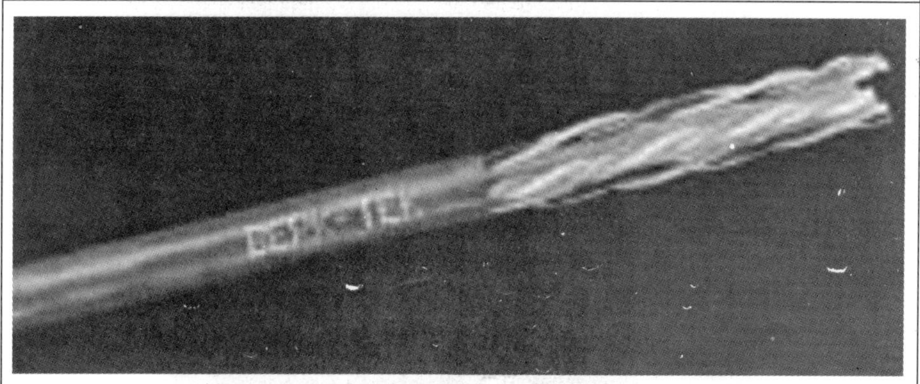

PolyOne first functionalizes the nanoclay with a surfactant and then adds caprolactam monomer. The caprolactam is polymerized into nylon 6, which leaves the nanoclay particles dispersed without the work and expense of melt blending, and the nylon is bound to the clay by its affinity for the surfactant chains. This "alloy" technology, licensed from Toyota Corp. of Japan, reportedly achieves better properties than melt blending and uses half as much nanoclay because the alloy doesn't reagglomerate.

The loss of properties from reagglomeration "turned a lot of people off when they first tried nanoclay materials," says Roger Avakian, chief technology officer of PolyOne. "They had paid a lot of money for the nanoclay and got worse properties than before. When we gave this new Nanoblend to customers, we didn't call it a nanocomposite because that had got such a bad reputation. We just said, 'These are property-enhancing.'"

Nylon Nanoblends for injection molding are available with 2.5% and 8% nanoclay, the latter being a high level for a precompounded material. Nylon Nanoblend with 2.5% nanoclay shows better mechanical properties than a melt blend of 5% nanoclay (see table). In late 2004, PolyOne commercialized its first alloyed nylon Nanoblend in exterior trim for two car programs.

Nanoclays improve UL ratings in wire and cable by eliminating dripping and promoting char formation at relatively low cost. Sud-Chemie's nanoclay in polyolefin cable jacket formulation at right does not drip, while unfilled sample at left drips

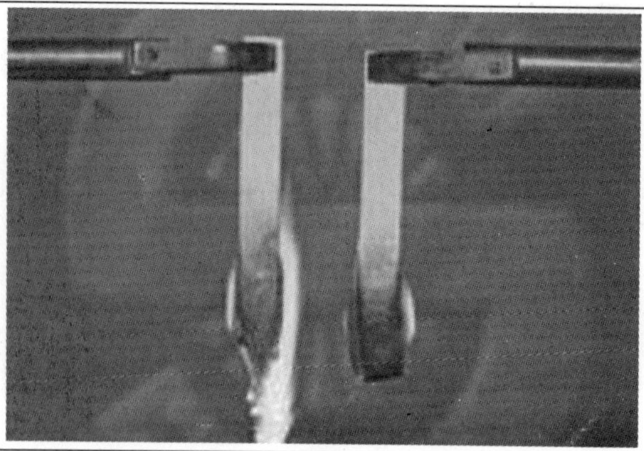

Less Costly Nano-Carbon

While carbon nanotubes are considered promising as electrically conductive nano-reinforcements, they are currently hobbled by high cost—up to $550 lb or more. While less expensive methods of producing carbon nanotubes are being developed by Bayer MaterialScience and others, other electrically conductive nano-particles are being introduced as less expensive substitutes for carbon nanotubes. For example, Pyrograf Products makes vapor-grown Pyrograf III carbon nano-fibers that cost around $100 to $150/lb, which is low enough to attract some compounders. Premix Thermoplastics has a new Pre-Elec line of electrically conductive masterbatches containing Pyrograf's nanofibers in a range of resins from PP to PEEK.

Nanosperse, a two-year-old nano-compounder, has developed concentrates with Pyrograf nano-fibers in TPU, epoxy, and thermoset vinyl ester. It uses patented dispersion technology initially developed for the US Air Force at Wright-Patterson AFB and licensed from the University of Dayton Research Institute in Dayton, Ohio. Concentrates are available with up to 20% nano-fiber.

PolyOne's in-situ polymerized Nanoblend nylon 6 chemically ties the resin to the nanoclay in the reactor. It uses half as much nanoclay to achieve better properties than melt blends (See Table)

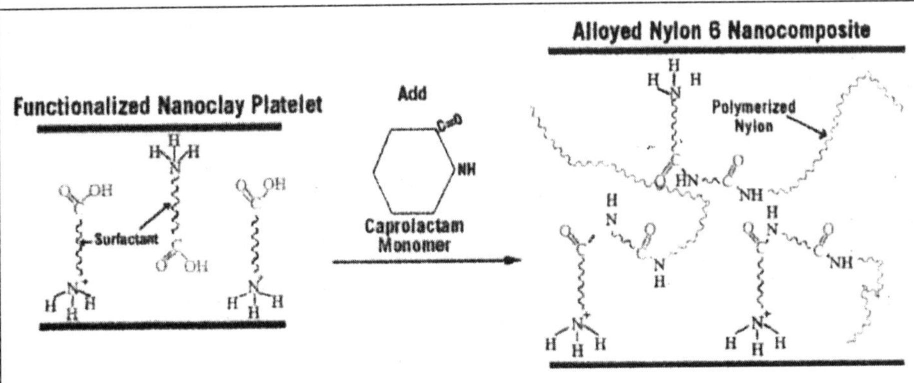

More New Nano-Additives

Naturalnano Inc., a two-year-old development company, presented at the San Francisco conference recently discovered, naturally occurring alumina silicate

Enhanced Properties of In-Situ Nanoclay Alloys			
Property	**Neat Nylon**	**Melt Blend**	**In-Situ Alloy**
Clay %	0	5	2.5
Flex Str., kpsi	17.4	19.7	22
Flex. Mod., kpsi	454	604	645
Tens. Mod., kpsi	435	575	623
Notched Izod, ft-lb/in.	0.6	0.6	0.7
Source: PolyOne.			

nanotubes called halloysite. Naturalnano has patents pending on the process to extract and classify halloysite tubes, which range from 1 to 40 microns long and have inner diameters of 20 to 50 nm.

DDG Cryogenics, a new firm spun off from Boeing in March 2005, is commercializing new cryogenically milled metal, ceramic, and composite nano-powders. Initial applications are in aluminum metal-matrix composites, but DDG's relatively inexpensive batch process is also suited to plastics. DDG has sampled its nano-powders for two years in nylon 6 and 12 applications.

Sud-Chemie reports that it has ground its nanoclays into smaller sizes that offer better dispersability. Its newer Nanofil 2, 5, and 9 reportedly disperse better than its older Nanofil 15 and 32. Its new Nanofil SE grades reportedly also disperse better in polyolefins, even resins without maleic anhydride modification.

Dyneon LLC is exploring block copolymers with controlled architecture to improve exfoliation of nanoclay platelets in melt compounding. Block copolymers with different levels of hydrophobicity can help disperse hydrophilic nanoclays in hydrophobic resins. The block copolymers are functionalized with amine, epoxy, anhydride, and acid groups.

Nanoclay in PVC

Elam EL Industries of Israel, which makes electroluminescent wires, is using nano-particles of zinc oxide (from Degussa in Germany) and titanium dioxide (from Sachtleben Chemie GmbH in Germany) in acrylic masterbatches for uv protection of fluorescent dyes in PVC and PVDF compounds. Conventional uv stabilizers reduce photo-luminescence, particularly in fluoropolymers. Elam's nano-additives don't actually prevent uv degradation of fluorescent dyes, but they keep the dyes from migrating to the surface so they don't degrade.

PolyOne'S PP-based Nanoblend LST compound in this material-handling tray replaced a PPO alloy, giving better dimensional stability and cutting the cycle time by 30 sec.

Meanwhile, the University of Belfast, Northern Ireland, reports the development of a nanocomposite plasticizer for PVC to replace DEHP plasticizer, which is banned in some markets because of concerns over chemical leaching. The nano-plasticizer is a soft EVA-carbon monoxide terpolymer blended with 0.5% to 4% of synthetic fluorohectorite nanoclay. The clay is sold as Somasif ME from Co-Op Chemical Co. in Japan.

Surprisingly, the university researchers found that adding nanoclay in EVA-CO to PVC actually made flexible PVC clearer. Nanoclays are smaller than the wavelength of visible light, so they typically have no optical effect. But PVC with 2% nanoclay transmits more light than PVC without clay—and more light than EVA-CO with clay by itself.

Researchers found that compounding in two stages was gentler to the PVC and protects it from contact with the fluoro-clay. First they blended 60% PVC with 40% nano-filled EVA-CO, which melts at a lower temperature than PVC. The EVA-CO coats the clay, so it doesn't come in contact with the PVC. Then 20% more PVC is added a quarter of the way down the extruder barrel.

For Lab Testing of tiny Samples, DSM Xplore Developed a Twin-Screw Batch Extruder that uses Inserts to Change Batch Size

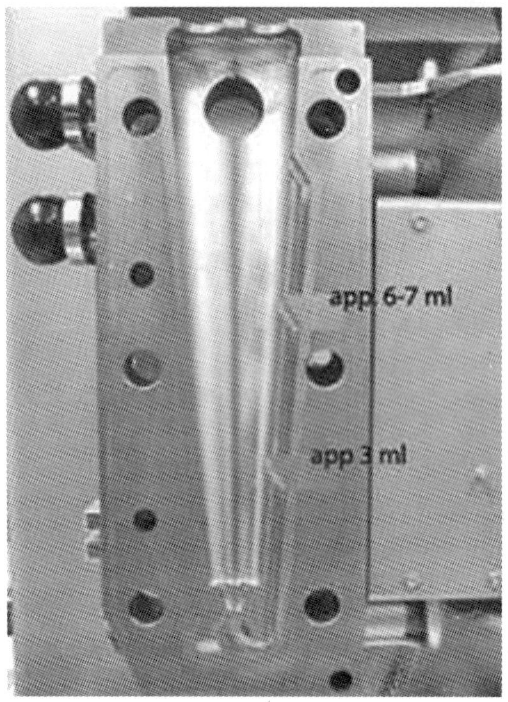

News in Nano Testing

DSM Xplore in Geleen, The Netherlands, represented in the U.S. by Dieter Scientific, presented an unusual laboratory device for developing expensive products like nanocomposites. It is a tiny conical twin-screw extruder designed for batch operation, unlike a normal continuous-feed extruder. It is fed by a syringe-like device. A two-position valve either discharges melt at the end of the screws or recirculates it back to the feed end of the screws in a closed loop.

DSM Xplore, which originally built lab equipment for materials supplier DSM, became a separate unit two years ago.

DSM's original lab extruder processed 5-cc batches, so it needed only a pinch of expensive nano-materials to run tests. The latest generation uses inserts to change the effective barrel length and batch size from a minimum of 2.5 cc up to 15 cc. DSM has sold over 50 of the lab units. Prices range from about $80,000 to $100,000.

Elementis Specialties, a supplier of montmorillonite nanoclay, has developed a new infrared spectroscopy test to assess platelet dispersion and orientation. Based on IR absorption by the silicon-oxygen bonds in the clay, a broad absorption band indicates agglomerated clay. A band that narrows and forms a peak indicates the clay is exfoliated.

Elementis's method can also measure platelet orientation, which is key to barrier properties in extruded or molded parts. A sample is rotated around the Y axis of an IR beam while the beam is successively pointed in Y and Z directions. If the IR absorption band varies in intensity, it indicates non-random orientation of the platelets. Alternative test methods like electron microscopes and X-ray diffraction analysis are much more complex and provide less clear results, Elementis says.

(Jan H Schut is a Senior Editor, Plastics Technology.)

7

Nanobiotechnology in North Carolina

Sarah Jackson, Maria Rapoza, Rudy Juliano, Kenneth Gonsalves and Ken Tindall

North Carolina is in a position to be profoundly affected by advances in nanobiotechnology, the field of nanotechnology specializing in creating novel nanoscale materials with desired biological, physical and chemical properties. The state has a strong research base, with nearly $100 million in nanotechnology research already funded. Current research in nanobiotechnology portends tomorrow's advances in drug delivery, medical imaging and diagnostics, and even computer circuits.

Nanobiotechnology may have broad impacts across many sectors that are well-established in the state, including health care, textiles and electronics. Coordination of statewide research activities and development of infrastructure that serves industrial development will ensure a stronger role for North Carolina in nanotechnology, bolster existing industries, and bring new jobs to the state.

This report describes nanotechnology, nanobiotechnology and some of their applications, with a focus on North Carolina's leadership role in nanobiotechnology today and in the years to come.

What is Nanotechnology?

Nanotechnology is the application of techniques to manipulate and study matter at the nanoscale – a world smaller than mere miniaturization, typically involving particles in the range of one billionth to 100 billionths of a meter in size. That's about 1/80,000 the thickness of a human hair. At this level, carefully arranged atoms and molecules in various materials have been found to have new and often unexpected properties. Drawing on the convergence of physics, chemistry, materials science and engineering, nanotechnology allows for the creation of novel materials and products with desired properties such as increased strength or better conductivity. The ability to design nanomaterials is the beginning to have broad impacts on many fields including chemistry, medicine, engineering, and computational science.

What is Nanobiotechnology?

Nanobiotechnology represents the convergence of nanotechnology and biotechnology, yielding materials and products that use biological molecules in their construction and are designed to affect biological systems. Several applications of nanobiotechnology include:

- Engineering biomolecules for non-biological use, such as DNA-based computer circuits using nanotechnology tools such as medical diagnostic devices and medical imaging to study biology
- Combining nanomaterials with biological systems for outcomes such as targeted drug therapies.

The state of Nanobiotechnology in North Carolina

North Carolina is already taking part in cutting-edge nanobiotechnology research. The nanotechnology magazine *Small Times* in 2005 ranked the state among the top 10 regions in the country for nanotechnology research, citing strong academic nanotechnology programs and a high concentration of nanotech researchers and graduate students. The state has at least 27 university-based nanotechnology centers and institutes. The University of North Carolina-Chapel Hill was named one of the top 10 nanotechnology universities in the United States in a 2006 report by the Southern Growth Policies Board. North Carolina State University and UNC-Chapel Hill both ranked in the top 10 in industrial outreach in nanotechnology.

One measure of the strength of the university research programs is the level of nanotechnology funding. Between 1995 and 2004, the National Science Foundation awarded 139 nanotechnology grants to North Carolina researchers, totaling more $53 million, according to the Southern Growth Policies Board. NCSU, UNC-Chapel Hill and Duke University rank in the top 100 institutions nationwide based on the total funding awarded in nanotechnology.

In 2005, UNC-Chapel Hill received a $3.8 million grant from the National Cancer Institute to establish the Carolina Center of Cancer Nanotechnology Excellence. Led by principal investigator Dr. Rudy Juliano, the center is one of seven in the nation focused on applying nanotechnology to improve cancer diagnostics, imaging and therapy. Dr. Shelton Earp, director of the UNC-Chapel Hill Lineberger Cancer Center, said, "Our inclusion in this National Cancer Institute program with some of the world's premier physical science universities is a tribute to UNC's ability to put together teams across academic boundaries. This synthesis of engineering, material science and medicine is both the future of patient care and attractive to industry at all levels."

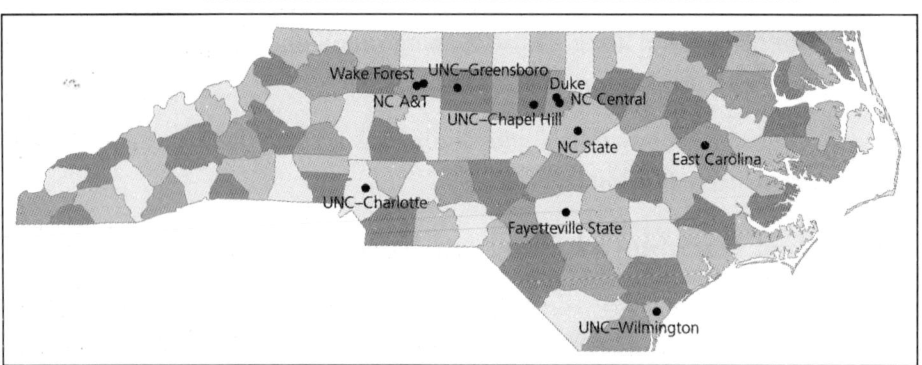

Funding in Nanotechnology Research at North Carolina Universities 2000-2005	
Institution	**Funding**
Duke University	$18,756,011
East carolina University	$35,000
Fayetteville State University	$91,896
North carolina A&T State University	$10,983,115
North carolina Central University	$2,317,474
North carolina State University	$33,266,273
University of north carolina – Chapel Hill	$25,027,505
University of north carolina – Charlotte	$5,421,875
University of north carolina – Greensboro	$44,999
University of north carolina – Wilmington	$498,083
Wake Forest University	$2,362,917
Total	**$98,805,148**

The University of North Carolina-Greensboro and North Carolina A&T State University are building a Joint School of Nanoscience and Nanoengineering to train graduate students in basic and applied nanotechnology research. The universities have secured funding for a $10 million core laboratory building. Dr. Rosemary Wander, associate provost at UNC-Greensboro, lauded the research endeavor as "a unique effort by two campuses that has the potential to dramatically impact North Carolina'seconomy and position the state as a major international player in this rapidly developing area."

Another way of assessing North Carolina's impact in nanotechnology is to examine scientific publications from researchers in the state. Between 1995 and 2004, North Carolina researchers published 385 articles on nanomedicine and nanobiology – categories within the field of nanobiotechnology – and more than 1,000 publications in the broader field of nanotechnology. Top-notch nanotechnology research is being conducted at institutions throughout the state including BD Technologies, Duke University, Carolinas Medical Center, East Carolina University, Fayetteville State University, GlaxoSmithKline, the National Institute of Environmental Health Sciences, North Carolina A&T State University, North Carolina Central University,

N.C. State University, RTI International, Shaw University, UNC-Chapel Hill, UNC-Charlotte, UNC-Greensboro, UNC-Wilmington, and Wake Forest University.

From talented researchers, new innovations may lead to new commercialization possibilities. UNC-Chapel Hill, Duke, and NCSU already hold several nanotechnology patents, as do North Carolina companies GlaxoSmithKline and Trimeris. More than 250 nanotechnology patents were issued to North Carolina assignees between 2003 and 2005, according to the US Patent and Trademark Office. Additionally, start-up nanobiotechnology companies in North Carolina are already working to develop a wide range of nanobiotechnology products.

Strengthening Nanotechnology in North Carolina

North carolina has long recognized that encouraging science and technology leads to economic prosperity. the governor's task Force on nanotechnology and north carolina's economy has developed a strategic roadmap to promote the development of nanotechnologies and help to bring new high-wage jobs in nanotechnology to north carolina. Members of the task force came from business, academia and the public sector.

"A Roadmap for nanotechnology in north carolina's 21st century economy" details their findings including a framework for increasing collaboration between industry and universities, further developing nanotechnology research centers at north carolina universities, improving education and workforce training, and educating north carolina citizens and policy makers on nanotechnology issues.

Dr. Robert McMahan, executive director of the NC office of Science and technology, described the roadmap as "a very clear call to action for the political, industrial, and university communities to build the nanotechnology and bionanotechnology community in the state."

Selected Nanobiotechnology Companies in North Carolina

Company	Location	Area of nanobiotechnology research
Advanced Liquid Logic	RTP	Lab-on-a-chip devices
Centice Corporation	Durham	Spectroscopy sensors for molecular recognition
HyperBranch Medical Technologies	Raleigh	Nanomaterials and devices for ocular surgery; drug delivery
LaamScience	RTP	Anti-viral surface coating using nanomaterials
Liquidia Technologies	Morrisville	Engineered nanomaterials for manufacturing processes in many industries

Contd...

Contd...		
NanoCor Therapeutics	Chapel Hill	Nanoparticles for delivering gene therapy for chronic heart failure treatment
NanoTechLabs	Yadkinville	Nanomaterials
PharmAgra Labs	Brevard	Nanomaterials for pharmaceutical and biotechnology.
Tiny Technology	Charlotte	Nanotechnology consulting and education
QuarTek International	Greensboro	Nanomaterials, devices and sensors
Xintek	RTP	Nanomaterial-based field emission technologies for X-ray imaging

Start-up companies have sprung from the strong base of nanotechnology research in North Carolina universities. The majority of nanotechnology companies are located in Research Triangle Park, the Piedmont Triad, or the Charlotte area near universities with major nanotechnology research programs. Nanobiotechnology is particularly strong in the Research Triangle region, while other regions are developing infrastructure to support research and industrial growth. With further development of its nanotechnology infrastructure, the state can position itself more favorably to nanotechnology businesses, tap its strong base of researchers and leverage its low overhead costs.

Nanotechnology clusters are developing around the nation's biotechnology "magnets," particularly in California and the Northeast. Many states already have a nanotechnology initiative in place, and are actively investing in research centers and promoting business development.

The State of North Carolina is well known for its visionary investments in biotechnology. These investments have been handsomely rewarded. Biotechnology companies in the state now employ more than 48,000 people statewide and account for a $3 billion annual payroll. North Carolina leaders are embracing nanotechnology development. Governor Mike Easley formed a Task Force on Nanotechnology to provide direction for advancing nanotechnology in North Carolina and to identify areas that need investments to ensure the state's future in the nanotechnology sector. North Carolina will build on its biotechnology, health care, and information technology sectors as nanotechnology increases in importance.

The Economics of Nanotechnology

In the last decade, global funding for nanotechnology research has reached $9 billion annually, according to the President's Council of Advisors on Science and Technology. The world market value for nanotech materials estimated to be $1 trillion by 2015, and some analysts say nanotechnology may account for the creation of 2 million jobs worldwide.

Recognizing the enormous potential impact of nanotechnology, the US established the National Nanotechnology Initiative in 2001 to fund nanotechnology research.

Nanobiotechnology Conference at the Biotechnology Center

The North Carolina Biotechnology center held its first North Carolina Nanobiotechnology Conference in 2006, drawing state policy makers, industry leaders and academic researchers from the Carolinas Medical Center, Duke University, East Carolina University, Fayetteville State University, North Carolina Central University, North Carolina State University, Shaw University, the University of North Carolina-Chapel Hill, the University of North Carolina-Charlotte, Wake Forest University, and Winston-Salem State University.

The conference gave north carolina researchers the opportunity to present and exchange information, from toxicity studies of nanoparticles to DNA-based computer chips and the latest research on creating and selecting desired nanomaterials.

Dr. Robert McMahan, the state's science and technology advisor, and executive director of the NC office of Science and technology, called the conference participants "the embodiment of the kind of innovative strategic thinking for which North Carolina is recognized internationally."

Between 2001 and 2005, the federal government invested more than $4 billion on nanotechnology research. Funding for nanotechnology research has been given a high priority. The Bush Administration requested $1.45 billion in the 2008 budget for the National Nanotechnology Initiative. In addition to federal funding, private investments and state spending account for about $2 billion in annual nanotechnology investments.

Top 10 Nanotechnology Patent-Holder Nations based on Patents Issued in 2005

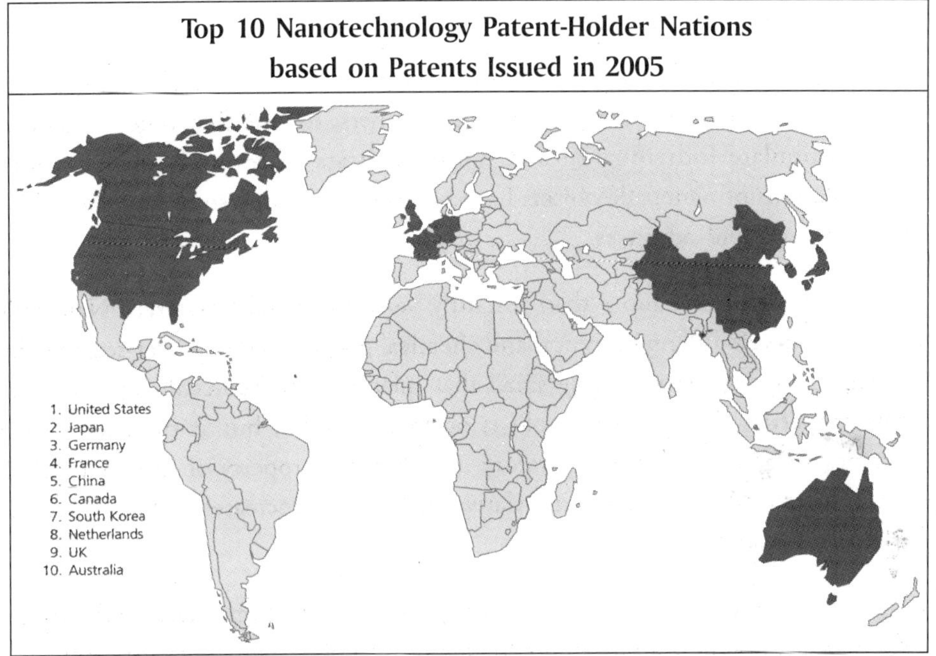

The nanotechnology field is sufficiently developed that discoveries are beginning to transition from scientific novelties to commercial applications. One measure of the commercialization of nanotechnology is the number of patents issued. Between 1995 and 2004, more than 8,600 nanotechnology-related patents were issued in the United States. The majority of nanotechnology-related patents worldwide are issued to assignees in the United States, though other nations, including Japan and Germany, play a major role in global nanotechnology research.

Nanotechnology-based products are already on the market in cosmetics and sunscreens, machinery coatings, and electronic displays, and new applications may impact medicine, pharmaceuticals, genomics and proteomics in the near future. The next wave of applications is expected to include many from the nanobiotechnology field. A technical advisory group for the President's Council of Advisors on Science and Technology predicted in May 2005 improved biological sensors and portable and convenient medical diagnostic devices will be available in the next five years. Furthermore, targeted drug therapies and enhanced medical imaging are expected within five to 10 years.

Understanding the Science

The field of nanotechnology took form in the early 1980s, when the development of scanning probe microscopy allowed researchers, for the first time, to observe and manipulate individual atoms. Through years of research and advances, manipulations of nanoscale objects have become less cumbersome and the field is now making rapid advances.

Nanomaterials are more than just tiny versions of bulk material. Nanoscale materials possess different physical and chemical properties than larger-scale matter because their size is sufficiently small that quantum mechanics dictates some of their properties. For example, gold is normally a solid, but at the nanoscale it becomes a liquid at room temperature. Another key property of nanomaterials is that as a particle gets smaller, its relative surface area increases and its electronic structure changes.

The unique properties of nanomaterials due to quantum effects and surface-area effects allow new applications for many materials. The ability to create and design novel materials is ultimately why nanotechnology promises to make a huge impact on so many fields.

Nanomaterials can be constructed by a top-down approach or a bottom-up approach. In the top-down approach, the nano builder reduces the size of a material of interest until it reaches nanoscale proportions. In the bottom-up approach, nanostructures are built atom by atom or molecule by molecule by either manipulating individual atoms or creating conditions where components can self-assemble. Self-assembly is much faster than building nanomaterials atom by atom; engineering new nanomaterials that self-assemble may provide a critical manufacturing solution. The field of biology provides many examples of self-assembling molecules, such as proteins that fold into a specific configuration. Studies of naturally occurring self-assembly may lend insight to designing new nanomaterials.

Designing nanomaterials differs from traditional chemistry because nanotechnology allows unequaled molecularlevel control over materials. Novel nanomaterials are being developed for biotechnology and medical applications. And the marriage of nanotechnology and biotechnology has led to the emerging field of nanobiotechnology.

Research in nanotechnology, biotechnology, and information technology intersects increasingly. Multidisciplinary research involving medical researchers, biologists, chemists, physicists, materials scientists, and engineers is propelling nanobiotechnology forward. The convergence of these once-distinct fields is the essence of nanobiotechnology.

Because nanomaterials are of the same scale as biological molecules, nanomaterials may open new possibilities for monitoring and intervening in biological systems. Researchers are using nanobiotechnology to create new contrast agents for cell imaging, to deliver gene therapy, and to analyze cellular processes. Nanomaterials are being investigated as a potential scaffold for cell transplantation in the treatment of diseases such as Parkinson's and diabetes.

NanoTechLabs

Nanotechnology may some day offer treatments for neurological conditions such as Parkinson's disease, epilepsy and depression. Nanotechlabs of yadkinville, N.C., has recently developed a new type of carbon nanotube electrode for electrical stimulation of the brain.

Implantable electrodes may some day be used to for deep brain stimulation. Analagous to a heart pacemaker, the device would emit electrical impulses to stimulate the brain.

Nanotechlabs and commercial research partners Foster-Miller and innerSea technology are developing nanomaterials that are biologically compatible and readily manufactured. With further research, new treatment options for neurological diseases may be possible.

Nanotechnology may also play an important role in developing tissue-engineering materials that are more compatible with the body. On-site diagnostic work to assess the presence or activity of a particular molecule can be faster and more sensitive using nanoscale tags. Drug delivery systems that use nanoparticles to carry drug molecules only to the desired location could reduce side effects and increase potency.

Interfacing nanomaterials with biomolecules may allow a wide variety of medical applications as well as analytical tools for advancing genomics – the study of all the genes in living organism – and proteomics – the study of all the proteins in a living organism.

Nanoparticles and Gene Therapy

Could nanotechnology hold the key to unlocking gene therapy for diseases such as muscular dystrophy? Researchers at the University of north carolina-chapel hill are working on the next generation of gene therapy using biological nanoparticles.

For years, researchers have been studying gene therapy as a tool to treat diseases that are caused by the absence of a key protein in the body. By introducing the correct gene into cells, the therapy should enable the patient to make the missing protein. though gene therapy is a promising concept there have been problems with expressing sufficient quantities of protein and targeting the gene to desired tissues.

Samulski

UNC-Chapel hill Professor Jude Samuiski founded Asklepios Bio Pharmaceutical, inc. (AskBio) to develop and commercialize proprietary biological nanoparticles for efficient, safe therapy of diseases such as muscular dystrophy, hemophilia and congestive heart failure. The North Carolina Biotechnology center has provided $165,000 in funding for AskBio, including a 2005 pre-clinical study of biological nanoparticles in congestive heart failure treatment. The company is also conducting the first human clinical trials of a gene therapy for Duchenne Muscular Dystrophy.

"After years of encouraging preclinical results, I'm excited that AskBio will soon be able to bring this promising new therapy into the clinic, and look forward with a great deal of optimism to offer this initial step toward hope for the Duchenne Muscular Dystrophy community," Samulski told the Muscular Dystrophy Association.

Though the potential impact of nanobiotechnology is very real, there are still significant hurdles that must be overcome. There are technical problems of moving from lab scale to industrial scale. Better instruments for observing, analyzing, and positioning molecules are needed. Most significantly, though, issues surrounding the impact of nanomaterials on human health and the environment, and related ethical concerns, must be explored.

Environmental Safety and Human Health Concerns

As nanotechnology moves forward, it is possible that some nanomaterials may cause toxic effects on humans and in the environment. Because of their small size, many nanomaterials are readily absorbed by the body.

Computer Processors Using Self-assembling DNA Nanostructures

Cutting-edge research into the computers of tomorrow includes the use of DNA nanostructures in computer processors. Already, researchers have assembled DnA nanostructures for experimental use and are working toward practical applications.

Dr. Thomas LaBean, associate research professor of computer science and chemistry at Duke University, described a research project at Duke involving the use of DnA nanostructures to detect and modify faulty gene expression in a programmable manner that is analogous to a computer computation.

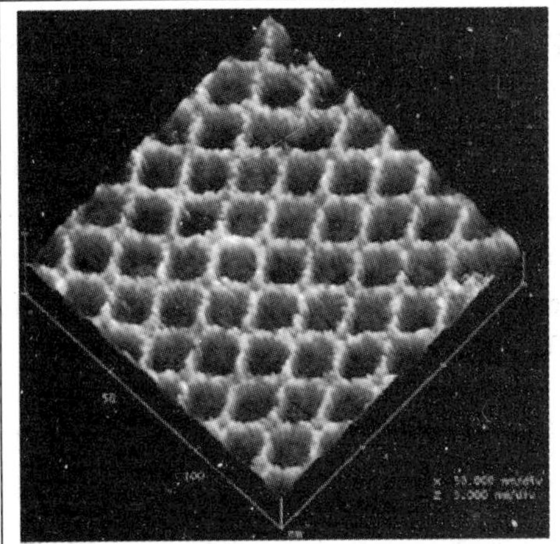

Surface plot of DNA nanogrid constructed by Duke DNA nanotech group

"A simple system has been demonstrated outside of cells," said LaBean, "but a system functioning inside cells is probably at least 10 years away." Another area of research is using DNA nanostructures as "smart glue" to direct the assembly of other materials that might have useful electronic properties or photonic properties. Photonics uses photons—the fundamental particles of light—to transfer or store information similar to the use of electrons in electricity.

Complex DNA nanostructures have been used to organize proteins, metals, semiconductor nanoparticles and carbon nanotubes. A reliable DNA-based assembly method for manufacturing electronic components or devices has not yet been demonstrated, according to LaBean, but "this direction could perhaps result in products within the next 10 years."

Though DNA nanostructure research is still in its infancy, this technology is promising because nature has proven DNA to be able to form well-organized, complex nanostructures that can exhibit diverse and extremely useful properties.

For example, researchers at the North Carolina State University Center for Chemical Toxicology Research and Pharmokinetics found that carbon nanotubes can be taken up by a human skin cell line and alter protein production. The skin represents a very likely route for exposure to nanomaterials, and these kinds of studies indicate that not all nanomaterials are benign.

However, nanomaterials should not be lumped into a single category. Different nanomaterials can have vastly different properties, and therefore very different toxicology. For example, some nanoparticles can be readily broken down by the body while others may remain in the body presenting different risks associated with long-term exposure. Additionally, for some nanomaterials, changes in manufacturing conditions can dramatically influence toxicity.

ASTM International, originally known as the American Society for Testing and Materials, is developing a "Standard Guide for Handling Unbound Engineered Nanoparticles in Occupational Settings" to provide guidance for controlling exposure to nanoparticles in the laboratory, particularly for nanoparticles that have undefined risks and exposure information.

In addition to addressing human health concerns, environmental toxicology research assessing the impact, routes of exposure and potential for bioaccumulation of nanomaterials will be critical to assess the ecological impact of these materials. More research needs to be done to answer questions surrounding the safety of different nanomaterials.

North Carolina researchers are poised to play an important role in these toxicity studies. Some of the nation's top toxicology programs are based in North Carolina including the National Toxicology Program of the National Institute of Environmental Health Sciences (NIEHS), the Environmental Protection Agency (EPA), the CIIT Centers for Health Research, in addition to strong programs in the state's universities. Clearly, a thorough understanding of the toxicology of nanomaterials, and optimization to reduce harmful effects, are essential as the field moves forward.

"To date, there are really only limited toxicity studies based on a very limited number or types of nanomaterials," said Dr. Amy Ringwood, an assistant professor of biology at UNC-Charlotte who uses oysters in her lab to study toxicity. "I regard it as important to consider toxicity issues in conjunction with the development of nanomaterials to understand potential problems. In doing so, we can work cooperatively and responsibly to develop safe nanotechnology strategies."

Ethical Issues

Another key area of concern is the public understanding of nanobiotechnology in our society. Emerging technologies such as genetically modified foods and stem cell research have challenged people to examine their beliefs about the natural order of the world. Nanobiotechnology will present new challenges.

Researchers at North Carolina State University are investigating public concerns about nanotechnology. A 2006 study of lay people who were provided reading materials about nanotechnology showed that 62 percent of participants had low trust in the ability of the government to manage risks associated with nanotechnology. Many participants cited past examples of new technologies that were poorly managed such as dioxin, PCBs, and asbestos. The most common concerns expressed about the use of nanotechnology included: creation of new weapons of mass destruction, undesirable military use or use of such weapons by terrorists; unknown long-term health risks of nanoparticles; and environmental problems such as new pollutants and unexpected effects on the environment.

Most participants did not fear that nanotechnology would lead to negative consequences as portrayed in the popular novel *Prey* by Michael Crichton. Nor were they concerned that nanotechnology would lead to self-replicating nanobots that consume living creatures and turn them into gray goo, as envisioned by Eric Drexler. Drexler, a scientist who advocates education about the benefits and risks of nanotechnology, says now that "fears of accidental runaway replication – loosely based on my 1986 gray-goo scenario – (are) quite obsolete." He adds that "all the hype diverts attention from more important issues – research directions, development paths, and the role of advanced nanotechnologies in medicine, the environment, the economy and in strategic competition."

As nanotechnology progresses, it will be important for the public to understand the technology and be involved in discussions about how nanobiotechnology will impact medicine, the environment, public health, and the economy.

Outreach efforts are being made to inform people about this emerging technology. The National Science Foundation recognized this need in 2005 when it awarded the largest grant ever made to the museum community to create a nationwide network of informal science centers, researchers and educators dedicated to engaging the public through a wide range of learning experiences about nanoscale science, engineering and technology. As one of nine museums in the NiSE Network, the Museum of Life and Science in Durham, NC, is offering a series of public forums on the social and ethical issues surrounding emerging nanotechnologies.

Involving North Carolina Researchers in the Ethical Challenges of Nanobiotechnology

In addition to engaging the public in understanding nanotechnology and potential risks, research scientists must also consider the ethical challenges that nanotechnology brings. At universities across the state, the next generation of researchers is being encouraged to address the ethics of nanotechnology research.

North carolina State University is the leading institution in a national project funded by the national Science Foundation to develop a curriculum in research ethics for doctoral candidates at land grant universities. Researchers from NCSU, North Carolina A&T State University, North Carolina Central University, Fayetteville State University and other institutions across the nation are contributing to this nanotechnology ethics model curriculum. All of the lead institutions began implementing the ethics curriculum in 2006-2007.

"Today's doctoral students are tomorrow's nanotechnology leaders," said the grant's project director, Dr. Gary Comstock, a professor of philosophy and director of the research and professional ethics program at NCSU. "Our project will involve them in rigorous discussion of the broader social and ethical dimensions of nanotech, preparing them better to discern and avoid potential risks and harms. I hope that the module will provoke careful and sustained reflection on the moral ramifications of nanotechnology."

Comstock

At Duke University, professors from the center for Biologically inspired Materials & Material Sciences and the Center for Biological Tissue Engineering wanted to go beyond the standard required ethics training for graduate students by challenging students to confront the ethical issues surrounding nanotechnology in an essay contest. "The range and breadth of the issues that they have identified are impressive," said Dr. Daniel Vallero, who led the effort. "They include both macro and microethical issues and they related directly to real-life and real-time research challenges."

"Engaging the public right now in discussing the risks and benefits of nanoscience research and its early applications is not only the right thing to do, it could help avert greater public concern and policy issues later on," said Barry Van Deman, president and CEO of the Museum.

Barry Van Deman

Summary

Nanobiotechnology is poised to make rapid advances in the near future. North Carolina is implementing strategies to strengthen the state's role in this field. By creating a strong nanobiotechnology cluster in the state, North Carolinians will reap the economic and social benefits of this growing endeavor.

UNC-Charlotte/Carolinas Medical Center Conference

The University of North Carolina at Charlotte and the Carolinas Medical center hosted more than 250 people at the 2005 nanotechnology conference "Nanoscale Science and Engineering: a convergence of top-Down and Bottom-up Approaches." the conference provided a forum for researchers to discuss national initiatives in nanotechnology, nanotechnology in medicine, and other emerging nanotechnology research.

In recent years, Unc-charlotte and the carolinas Medical center have created a strong partnership to develop applications of nanotechnology for biomedical research. Their collaborative efforts have yielded promising research, such as investigating the use of nanostructures to deliver gene therapy in the treatment of muscular dystrophy and to deliver antibiotics to osteoblasts for the treatment for bone infections.

Hosting a nanotechnology Conference in charlotte has been a natural extension of such active research in nanomedicine. The North Carolina Biotechnology center, Unc-charlotte, and the Carolinas Medical Center also are hosting the 2007 nanotechnology conference in Charlotte focused on biomedical nanostructures.

(Sarah Jackson, Maria Rapoza, PhD, Ken Tindall, PhD, North Carolina Biotechnology Center,

Rudy Juliano, PhD, Department of Pharmacology, School of Medicine, University of North Carolina – Chapel Hill,

Kenneth Gonsalves, PhD, Department of Chemistry, University of North Carolina – Charlotte).

Additional Resources for Nanotechnology and Nanobiotechnology Information

Nanotechnology reports	Web site
Southern Growth Policies Board. 2006. Connecting the Dots: Creating a Southern nanotechnology Network	*http://www.southern.org/pubs/ConnectDots/ConnectExecSumm.pdf* Full text available at the Biotechnology center library or for purchase at *http://www.southern. org/pubs/pubs.shtml*
Governor's Task Force on Nanotechnology and North Carolina's Economy. 2006. A Roadmap for nanotechnology in north Carolina's 21st Century Economy	*http://www.ncnanotechnology.com/*
President's council of Advisors on Science and technology. 2005. The National Nanotechnology Initiative at Five years: Assessment and Recommendations of the National nanotechnology Advisory Panel	*http://www.ostp.gov/pcast/PCASTreportFINALlores.pdf*
National Science and Technology Council. 2004. National Nanotechnology Initiative Strategic Plan.	*http://www.nano.gov/nni_Strategic_Plan_2004.pdf*
Government Web Sites	
National Nanotechnology Initiative	*http://www.nano.gov*
National Institutes of Health Roadmap on Nanomedicine	*http://nihroadmap.nih.gov/nanomedicine/index.*asp
Nanotechnology Characterization Lab of the National Cancer Institute	*http://ncl.cancer.gov/*
Environmental Protection Agency National Center for Environmental Research: Nanotechnology Activities	*http://es.epa.gov/ncer/nano/index.html*
US Food and Drug Administration—Nanotechnology Site	*http://www.fda.gov/nanotechnology/*
National Institute of Environmental Health Sciences – National Toxicology Program Nanotechnology Safety Initiative	*http://ntp.niehs.nih.gov/go/nanotech*

Conted...

Conted...

University & Non-Profit Organization Web Sites	
Unc-chapel Hill Nanoscale Science Research Group	*http://www.cs.unc.edu/Research/nano/aims.html*
Center for Biologically Inspired Materials & Material Systems	*http://www.cbimms.duke.edu/*
Center on Globalization, Governance & Competitiveness—A research affiliate of the Social Science Research institute of Duke University studying the societal impacts of nanotechnology	*http://www.cggc.duke.edu/projects/cns/cns.html*
NCSU—Land grant University Research ethics module on nanotechnology	*http://www.chass.ncsu.edu/langure/modules/nanotechnology.html*
Wake Forest University Nanomedicine	*http://www.wfu.edu/nanotech/WFnanomeds.html*
Nanoscience and Nanotechnology Research Center at Shaw University	*http://faculty.shawu.edu/karoui/1NNRC/HomeMyNNRC.htm*
North carolina A & T State University—Nanoscale Interdisciplinary Research Team	*http://nirt.ncat.edu/*
Foresight Nanotech institute—a non-profit organization formed to guide nanotechnology research, public policy and education	*http://www.foresight.org/*
Johns Hopkins institute for NanoBiotechnology	*http://inbt.jhu.edu/*
The center for Biological and Environmental nanotechnology	*http://cben.rice.edu/*
Ciit centers for Health Research	*http://www.ciit.org/*
Nanotechnology news, products, events and information	*http://www.nanotechweb.org/*
NSIE (Nanoscale informal Science education) network—Article on Nanomedicine	*http://www.nisenet.org/publicbeta/articles/nanomedicine/index.html*
Other Web Sites of Interest	
Small times—Web site/magazine focused on business information about nanotechnology	*http://www.smalltimes.com/*
ASTM International—a standards development organization for technical standards. for materials, products, systems, and services.	*http://www.astm.org*

Conted...

Conted...	
Research Journal Articles on Nanotechnology	
Halberstadt *et al.*, Combining cell therapy and nanotechnology. Expert opin. Biol. Ther. (2006) Vol 6 Number 10, Pages 971-981	
Nanotechnology: assessing the risks, *Nano Today*, Volume 1, issue 2, May 2006, Pages 22-33, Andrew D. Maynard	*http://www.sciencedirect.com/*
International nanotechnology development in 2003: Country, Institution, and technology field analysis based on USP to patent database, *Journal of nanoparticle Research*, Volume 6, 2006, Pages 325-354, Zan huang, hsinchun chen, Zhi-kai chen, Mihail c. Roco	*http://citeseer.ist.psu.edu/huang04international.html*
Nanotechnology: public concerns, reasoning and trust in government. Public Understanding Sci., Volume 15, 2006, Pages 221-241 Jane Macoubrie	*http://pus.sagepub.com/cgi/content/abstract/15/2/221* (abstract only)
Witzmann and Monteiro-Riviere. Multi-walled carbon nanotube exposure alters protein expression in human keratinocytes. nanomedicine, Volume 2, 2006, 158-168.	

8

Nanotechnology Applications in Cancer

Shuming Nie, Yun Xing, Gloria J Kim and Jonathan W Simons

Cancer nanotechnology is an interdisciplinary area of research in science, engineering and medicine with broad applications for molecular imaging, molecular diagnosis, and targeted therapy. The basic rationale is that nanometer-sized particles, such as semiconductor quantum dots and iron oxide nanocrystals, have optical, magnetic, or structural properties that are not available from molecules or bulk solids. When linked with tumor targeting ligands such as monoclonal antibodies, peptides or small molecules, these nanoparticles can be used to target tumor antigens (biomarkers) as well as tumor vasculatures with high affinity and specificity. In the mesoscopic size range of 5-100 nm diameter, nanoparticles also have large surface areas and functional groups for conjugating to multiple diagnostic (e.g., optical, radioisotopic, or magnetic) and therapeutic (e.g., anticancer) agents. Recent advances have led to bioaffinity nanoparticle probes for molecular and cellular imaging, targeted nanoparticle drugs for cancer therapy, and integrated nanodevices for early cancer detection and screening. These developments raise

exciting opportunities for personalized oncology in which genetic and protein biomarkers are used to diagnose and treat cancer based on the molecular profiles of individual patients.

Introduction

The Cancer Problem

Human cancer is a complex disease caused by genetic instability and accumulation of multiple molecular alterations[1,2]. Current diagnostic and prognostic classifications do not reflect the whole clinical heterogeneity of tumors and are insufficient to make predictions for successful treatment and patient outcome[3,4]. Most current anticancer agents do not greatly differentiate between cancerous and normal cells, leading to systemic toxicity and adverse effects. In addition, cancer is often diagnosed and treated too late, when the cancer cells have already invaded and metastasized into other parts of the body. At the time of clinical presentation, for example, more than 60% of patients with breast, lung, colon, prostate, and ovarian cancer have hidden or overt metastatic colonies[5]. At this stage, therapeutic modalities are limited in their effectiveness. Due to these problems, cancer has overtaken heart disease as the leading cause of death for adults in the United States [United States Cancer Statisics, Centers for Disease Control and Prevention (CDC) *http://www.cdc.gov/cancer/npcr/uscs*].

Current problems and unmet needs in translational oncology include (*a*) advanced technologies for tumor imaging and early detection, (*b*) new methods for accurate diagnosis and prognosis, (*c*) strategies to overcome the toxicity and adverse side effects of chemotherapy drugs, and (*d*) basic discovery in cancer biology leading to new knowledge for treating aggressive and lethal cancer phenotypes such as bone metastasis. Advances in these areas will form the major cornerstones for a future medical practice of personalized oncology in which cancer detection, diagnosis, and therapy are tailored to each individual's cology in which genetic/molecular markers are used to predict disease development, progression, and clinical outcomes.

Cancer Nanotechnology

Cancer nanotechnology is emerging as a new field of interdisciplinary research, cutting across the disciplines of biology, chemistry, engineering, and medicine, and is expected to lead to major advances in cancer detection, diagnosis, and treatment[6,7]. The basic rationale is that metal, semiconductor, and polymeric particles have novel optical, electronic, magnetic, and structural properties that are often not available from individual molecules or bulk solids[8-10]. Recent research has developed functional nanoparticles that are covalently linked to biological molecules such as pep-tides, proteins, nucleic acids, or small-molecule ligands[11-18]. Medical applications have also appeared, such as the use of superparamagnetic iron oxide nanoparticles as a contrast agent for lymph node prostate cancer detection[19] and the use of polymeric nanoparticles for targeted gene delivery to tumor vasculatures[20]. New technologies using metal and semiconductor nanoparticles are also under intense development for molecular profiling studies and multiplexed biological assays[21-25].

Cancer Biomarkers

Biomolecular markers or biomarkers include altered or mutant genes, RNAs, proteins, lipids, carbohydrates, and small metabolite molecules, and their altered expressions that are correlated with a biological behavior or a clinical outcome. Most cancer biomarkers are discovered by molecular profiling studies based on an association or correlation between a molecular signature and cancer behavior. In the cases of both breast and prostate cancer, a deadly step is the appearance of so-called lethal phenotypes, such as bone-metastatic, hormone-independent, and radiation-and chemotherapy-resistant phenotypes. It has been hypothesized that each of these aggressive behaviors or phenotypes could be understood and predicted by a defining set of biomarkers. By critically defining the interrelationships among these biomarkers, it could be possible to diagnose and prognosticate cancer based on a molecular profile, leading to personalized and predictive medicine. That is, a unique molecular profile can be used to predict its ability to survive and grow under androgen-deprived and hypoxia and metabolic stress conditions, and the potential of certain cancer cells to evade host immune surveillance.

One of the first molecular profiling studies was reported by Golub *et al.*,[26] who showed that gene expression patterns could classify tumors, yielding new insights into tumor pathology such as stage, grade, clinical course, and response to treatment. Gene expression studies of cell lines further revealed that the molecular signature of each tumor is a result of the combined tumoral, stromal, and inflammatory factors of the original heterogeneous tumor[27]. The first clinical correlation of gene expression patterns with clinical outcome was reported for diffuse large B-cell lymphoma[28], a clinically heterogeneous disease. Whereas most (60%) of the patients succumbed to the disease, the remainder responded well to therapy and had prolonged survival. This variability in disease progression was correlated with a distinct pattern of gene expression. The concept of a specific molecular portrait for each later validated by Perou *et al.*,[29] and Bittner *et al.*,[30].

Most recent work on cancer molecular profiling by Rubin, Chinnaiyan, and their coworkers has combined cDNA microarrays with tissue microarrays for biomarker discovery and immunohistochemical validation[31-38]. For prostate cancer, a number of gene and protein biomarkers have been identified, including p504S (α-methylacyl coenzyme A racemase or AMAC, an enzyme involved β-oxidation of fatty acids), hepsin (HPN, a transmembrane serine protease), Pim-1, protease/KLK4, prostein, EHZ 2, and STEAP[39,40]. These markers appear to be excellent indicators of aggressive cancer behavior, such as metastasis and androgen independence.

Personalized Oncology

For applications in individualized therapy, biomarkers enable the characterization of patient populations and quantification of the extent to which new drugs reach their intended targets[41,42]. One example is the drug trastuzumab (Herceptin, Genentech/Roche), a monoclonal antibody designed to target amplified and overex-pressed *ERBB2* (also known as HER2) tyrosine kinase receptor found in ~25%–30% of breast cancers. FDA approval of trastuzumab was predicated on the availability of a test to detect *ERBB2* overexpression. Both an immunohistochemistry assay for the expressed protein (HercepTest, Dako) and a nucleic acid–based fluorescence in situ hybridization (FISH) test (PathVysion, Abbott) have been approved as in vitro diagnostics to guide trastuzumab treatment decisions. In another example, the clinical response of lung cancer patients to the epidermal growth factor receptor

(EGFR) tyrosine kinase inhibitor gefitinib (Iressa™, AstraZeneca) is associated with a small number of genetic mutations[43,44]. Thus, a molecular diagnostic test could be used to identify patients that are most likely to respond to this drug.

Despite these advances, critical studies that can clearly link biomarkers with cancer behavior remain a significant challenge. One difficulty is that most cancer tumors (especially prostate and breast cancer) are highly heterogeneous, containing a mixture of benign, cancerous, and stromal cells. Current technologies for molecular profiling, including RT-PCR, gene chips, protein chips, two-dimensional (2-D) gel electrophoresis, and biomolecular mass spectrometry (e.g., MALDI-MS, ES-MS, and SELDI-MS), are not designed to handle this type of heterogeneous sample[45,46]. Furthermore, a limitation shared by all these technologies is that they require destructive preparation of cells or tissue specimens into a homogeneous solution, leading to a loss of valuable 3-D cellular and tissue morphological information associated with the original tumor. The development of nanotechnology, especially bioconjugated nanoparticles, provides an essential link by which biomarkers could be functionally correlated with cancer behavior. Figure 1

Figure 1: Schematic diagram showing nanotechnology applications in cancer through molecular tumor imaging, early detection, molecular diagnosis, targeted therapy, and cancer bioinformatics

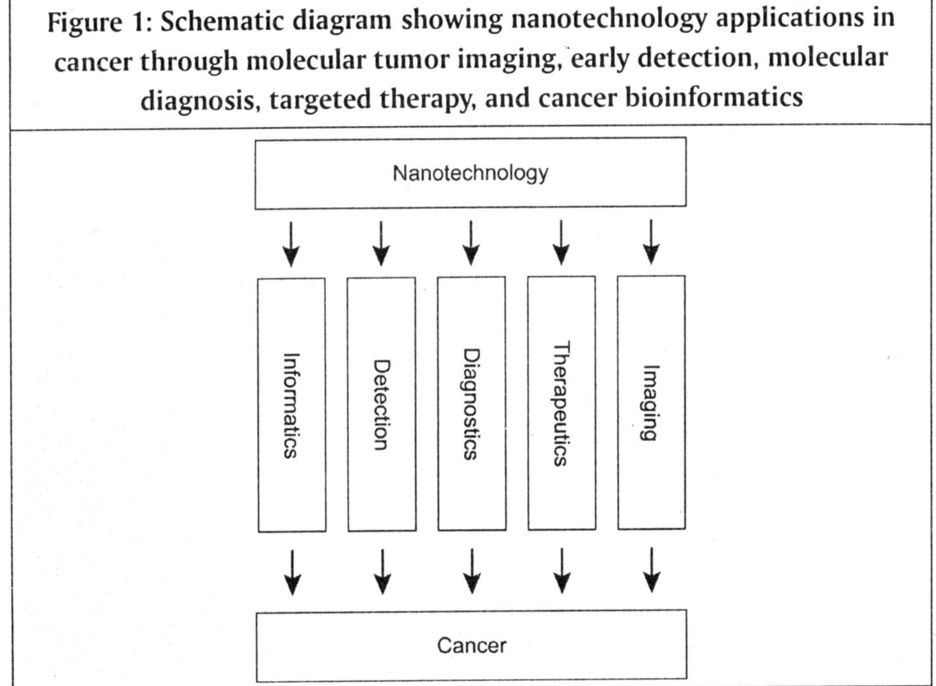

illustrates nanotechnology applications in cancer through molecular imaging, diagnosis, early detection, targeted therapy, and cancer bioinformatics. In the following, we describe the design and development of nanoparticle probes and their applications in cancer.

Nanoparticle Probes

A prototype nanoparticle is semiconductor quantum dots (QDs), tiny light-emitting particles on the nanometer scale that are emerging as a new class of fluorescent probes for in vivo biomolecular and cellular imaging (11-18) (Figure 2). In comparison with organic dyes and fluorescent proteins, QDs have unique optical and electronic properties. QDs have molar extinction coefficients that are 10-50 times larger than that of organic dyes, which make them much brighter in photon-limited in vivo conditions. Further, QDs emission wavelengths are size-tunable. For example, CdSe/Zns QDs of approximately 2 nm in diameter produce a blue emission, whereas QDs approximately 7 nm in diameter emit red light[47]. In recent work, researchers have pushed the emission wavelength into the near

Figure 2: Semiconductor quantum dots with quantum confinement and size-tunable optical properties. This image shows ten distinguishable emission colors of ZnS-capped CdSe quantum dots excited with a near-UV lamp. From left to right (blue to red), the emission maxima are located at 443, 473, 481, 500, 518, 543, 565, 587, 610, and 655 nm

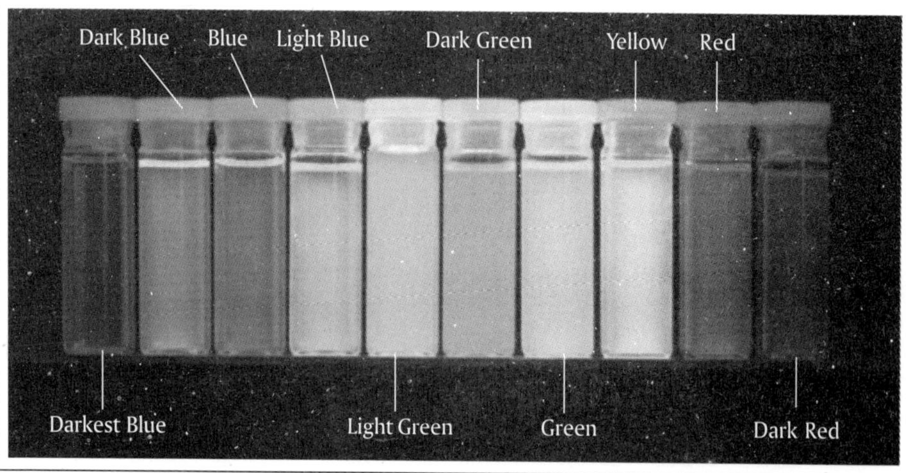

infrared (650 nm to 950 nm), to take advantage of improved tissue penetration depth and reduced background fluorescence at these wavelengths[48]. A key property for in vivo imaging is the broad QD Stokes shift, which can be as large as 300–400 nm, depending on the wavelength of the excitation light[49]. In conjunction with broadband absorption and narrow emission peaks of QDs, this property allows multiplexed imaging applications in which one light source is used to simultaneously excite multicolor QDs without the need for complicated instrumentation. Another important feature is the long-term photostability of QD imaging probes, which opens the possibility of investigating the dynamics of cellular processes over time, such as continuously tracking cell migration, differentiation, and metastasis. These properties have made QDs a topic of intensive interest in cancer biology, molecular imaging, and molecular profiling.

Dual-Modality Probes

Optical imaging is highly sensitive, but its applications in vivo and in human are hampered by a limited penetration depth in tissue and the lack of anatomic resolution and spatial information. Although near-infrared wavelengths can be used to improve the penetration depth, and 3-D fluorescence tomography can be used to provide spatial information[50,51], other imaging modalities, such as Magnetic Resonance Imaging (MRI), are much better for tomography and 3-D imaging. Thus, there has been considerable interest in developing dual-modality contrast agents for combined optical and MRI, which has exceptional tissue contrast and spatial resolution and has been widely used in the clinical setting. For example, by reacting superparamagnetic iron oxide nanoparticles with the fluorescent dye Cy5.5, Josephson and coworkers[52] have developed dual magneto-optical probes that are able to bind to apoptotic cells and are detectable by both fluorescence and MRI. Similarly, dual magnetic and optical imaging probes have been used to yield highly detailed anatomic and molecular information in living organisms[53]. These probes are prepared by conjugation of peptides to cross-linked iron oxide amine (amino-CLIO), either by a disulfide linkage or a thioether linker, followed by the attachment of the dye Cy5 or Cy7. Fluorescence quenching of the attached fluorochrome occurs by interaction with the iron oxide core, and also by electronic coupling among the dye chromophores (self-quenching). This class of dual-modality probes provides the basis for "smart" nanoparticles, capable of pinpointing their position through their magnetic properties, while providing information on their environment by optical imaging.

Recent research has shown that QDs can be linked with Fe_2O_3 and FePt to generate dual-function nanoparticles[(54,55)]. Others have entrapped Gd on the QD surface using polymer-conjugated lipids form dual-modality probes, but it is not clear whether these types of "hetero" nanostructures would be useful for in vivo medical imaging[(56,57)]. Research in our own group has created a new class of dual-modality nanoparticles by attaching a cluster of paramagnetic gadolinium chelates to polymer-coated QDs. Preliminary cellular and *in vivo* animal studies demonstrated that this class of nanoparticle is biocompatible and detectable by both fluorescence and MRI. In comparison with previous work, the polymer-protected QDs offer excellent optical properties (high-fluorescence quantum yields, narrow spectral widths, and high photostability), and the attached Gd chelates lead to significant T1 contrast enhancement with a brightening effect in MRI, as opposed to the T2 contrast with a darkening effect offered by iron oxide-based contrast. By linking to targeting lig-ands through a biocompatible polyethylene glycol (PEG) spacer, these dual-modality nanoparticle probes are promising for in vivo tumor imaging in animal models.

Multifunctional Platforms

Nanoparticles also offer a wide range of surface functional groups allowing chemical conjugation to multiple diagnostic and therapeutic agents. It is thus possible to design and develop multifunctional nanostructures that could be used for simultaneous tumor imaging and treatment, a major goal in cancer research and development. However, progress has been slow, and promising multifunctional platforms, such as dendrimers, liposomes, and PEBBLES (Probes Encapsulated in Biologically Localized Embedding), have not been able to deliver diagnostic and therapeutic agents to tumors in a selective and efficient manner[(58-62)]. Most of these studies are still at an early or proof-of-concept stage using cultured cancer cells, which are not immediately relevant to in vivo imaging and treatment of solid tumors.

Molecular Cancer Imaging

In comparison with traditional in vivo imaging probes or contrast agents [such as radioactive small molecules in positron emission tomography (PET) and single photon emission computed tomography (SPECT), gadolinium compounds in

MRI, and labeled antibodies], targeted QDs and other bioengineered nanoparticles provide several unique features and capabilities. First, their size-dependent optical and electronic properties can be tuned continuously by changing the particle size. This size effect provides a broad range of nanoparticles for simultaneous detection of multiple cancer biomarkers. Second, nanoparticles have more surface area to accommodate a large number or different types of functional groups that can be linked with multiple diagnostic (e.g., radioisotopic or magnetic) and therapeutic (e.g., anticancer) agents. This creates the opportunity to design multifunctional "smart" nanoparticles for multimodality imaging as well as for integrated imaging and therapy. Third, extensive research has shown that nanoparticles in the size range of 10-100 nm are accumulated preferentially at tumor sites through an effect called enhanced permeability and retention (EPR)[(63-66)]. This effect is believed to arise from two factors: (*a*) growing tumors produce vascular endothelial growth factors (VEGFs) that promote angiogenesis and (*b*) many tumors lack an effective lymphatic drainage system, which leads to subsequent macromolecule or nanoparticle accumulation. This causes tumor-associated neovasculatures to be highly permeable, allowing the leakage of circulating macromolecules and nanoparticles into the tumor interstitium.

Mapping Sentinel Lymph Nodes and Tumor Angiogenesis

In vivo imaging with QDs has been reported for lymph node mapping, blood pool imaging, and angiogenic vessels and cell subtype isolation. Ballou and coworkers[(67)] injected PEG-coated QDs into the mouse blood stream and studied how the surface coating would affect their circulation time. In contrast to small organic dyes (which are eliminated from circulation within minutes after injection), PEG-coated QDs were found to stay in blood circulation for an extended period of time (half-life more than 3 h). This long-circulating feature can be explained by the unique structural properties of QD nanoparticles. PEG-coated QDs are in an intermediate size range—they are small and hydrophilic enough to slow down opsonization and reticuloendothelial uptake, but they are large enough to avoid renal filtration. Webb and coworkers took advantage of this property and reported the use of QDs and two-photon excitation to image small blood vessels[(67,68)]. They found that the two-photon absorption cross sections of QDs are—two to three orders of magnitude larger than that of traditional organic fluorophores. Most recently, Jain and coworkers have used of QDs and QD-doped silica beads for differentiating tumor vessels from perivascular cells and matrix[(69)]. The results demonstrated a much clearer boundary between blood vessels and cells than that achieved by using traditional high-molecular-weight dextran.

For improved tissue penetration, Frangioni & Bawendi prepared a novel core-shell nanostructure called type II QDs[70], with fairly broad emission at 850 nm and a moderate quantum yield of ~13%. In contrast to the conventional QDs (type-I), the shell materials in type-II QDs have valence and conduction band energies lower than that of the core material. As a result, the electrons and holes are physically separated and the nanoparticles emit light at reduced energies (longer wavelengths). Their results showed rapid uptake of bare QDs into lymph nodes, and clear imaging and delineation of sentinel nodes (which are often surgically removed in patients diagnosed with breast cancer). This work points to the possibility that QD probes could be used for real-time intraoperative optical imaging, providing an in situ visual guide so that a surgeon could quickly and accurately locate and remove sentinel nodes or even small lesions (e.g., metastatic tumors), which may be difficult to identify without image guidance.

Tumor Targeting and Imaging

Akerman *et al.*,[71] reported the use of QD–peptide conjugates to target tumor vasculatures, but the QD probes were not detected in living animals. Nonetheless, their in vitro histological results revealed that QDs homed to tumor vessels guided by the peptides and were able to escape clearance by the reticuloendothelial system (RES). Most recently, Gao *et al.*,[49] developed a new class of multifunctional probes for simultaneous targeting and imaging of tumors in live animals. This class of QD conjugates contains an amphiphilic triblock copolymer that provides protection to aggregation and degradation in vivo and functional groups for targeting ligands for tumor antigen recognition. Addition of multiple PEG molecules provides improved biocompatibility and blood retention time. The use of an ABC triblock copolymer has solved the problems of particle aggregation and fluorescence loss previously encountered for QDs stored in physiological buffer or injected into live animals[11-18,72]. Detailed studies were reported on the in vivo behaviors of QD probes, including biodistribution, nonspecific uptake, cellular toxicity, and pharmacokinetics. Under in vivo conditions, QD probes are delivered to tumors by both a passive targeting mechanism and an active targeting mechanism. In the passive mode, macromolecules and nanometer-sized particles are accumulated preferentially at tumor sites through the EPR effect[63,64,66,73]. For active tumor targeting, Gao *et al.*,[49] used antibody-conjugated QDs to target a prostate-specific cell surface antigen, PSMA. Previous research has identified PSMA as a cell surface marker for both prostate epithelial cells and neovascular endothelial cells[74]. PSMA has been selected as an attractive target for both imaging and therapeutic intervention of prostate cancer. Accumulation and retention of PSMA antibody at the site of tumor growth is the basis of radioimmunoscintigraphic scanning (e.g., ProstaScint scan) and targeted therapy for human prostate cancer metastasis[75].

Correlated Optical and X-Ray Imaging

By integrating with an X-ray imaging machine, recent work by Nie and workers has also achieved optical and structural imaging on the same animal models. Figure 3 shows correlated X-ray and fluorescence images of high-quality, deep-red

Figure 3: Combined X-ray and fluorescent imaging of near-infrared-emitting (700 nm) QD probes injected into the peritoneal cavity of a mouse. The near-infrared emission of these QDs is readily detectable above background. Images were captured sequentially with a Kodak *in vivo* Image Station, with 625 nm excitation for fluorescence

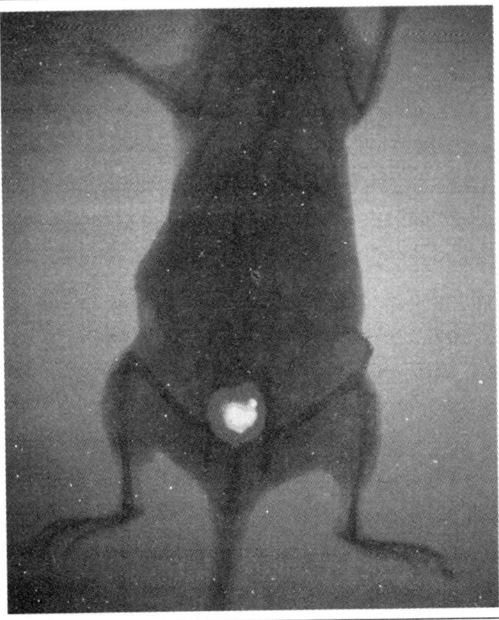

QDs injected into a mouse. The superimposed X-ray and optical images showed high sensitivity in detecting small tumors with low background and high signal levels in optical imaging, while providing detailed anatomic locations of small tumors in high-resolution X-ray. This type of correlated imaging combines the unique capabilities of different imaging modalities and is becoming increasingly utilized in basic and clinical cancer research. This powerful approach should provide new insights into cancer development, progression, and metastasis in animal models.

These studies using animal models have raised new possibilities for in vivo tumor imaging and have paved the way for further development of targeted tumor imaging in cancer patients. To develop clinical applications, the current nanoparticle probes encounter several challenges, such as limited tissue penetration, lack of spatial resolution in tumor depth and location, and potential toxicity concerns. Thus, there is an urgent need to develop broadly tunable near-infrared-emitting QDs to improve the tissue penetration depth. For clinical human applications, a major concern is likely the potential toxicity of QD probes, which has recently become a topic of considerable discussion and debate. Recent work by Derfus *et al.*,[76] indicates that CdSe QDs are highly toxic to cultured cells under UV illumination for extended periods of time. It has also been reported that the polymer-coated QDs could be toxic if significant aggregates are formed on the cell surface[77,78]. This is not surprising because the energy of UV-irradiation is close to that of covalent chemical bond and dissolves the semiconductor particles in a process known as photolysis, which releases toxic cadmium ions into the culture medium. In the absence of UV irradiation, QDs with a stable polymer coating have been found to be essentially nontoxic to cells and animals, with no observable effects on cell division and ATP production (D Stuart, X Gao, and S Nie, unpublished data). In vivo studies by Ballou and coworkers also confirmed the nontoxic nature of stably protected QDs[67]. Still, there is an urgent need to study the cellular toxicity, tissue and organ clearance, and in vivo degradation mechanisms of QD probes, as well as other nanoparticule formulations used for in vivo applications. For polymer-encapsulated QDs, chemical or enzymatic degradations of the semiconductor cores are unlikely to occur. But the polymer-protected QDs might be cleared from the body by slow filtration or excretion out of the body. This and other possible mechanisms must be carefully examined prior to any human applications in tumor or vascular imaging.

Molecular Cancer Diagnosis

Significant opportunities exist at the interface between biomarkers and nanotechnology for molecular cancer diagnosis. In particular, nanoparticle probes can be used to quantify a panel of biomarkers on intact cancer cells and tissue specimens, allowing a correlation of traditional histopathology and molecular signatures for the same material (see Figure 4). A single nanoparticle is large enough for conjugation to multiple ligands, leading to enhanced binding affinity and exquisite specificity through a multivalency effect. These features are especially important in the analysis of cancer biomarkers that are present at low concentrations or in small numbers of cells.

Figure 4: Schematic illustration of multiplexed detection and quantification of cancer biomarkers on intact cells or tissues with multicolor nanoparticle probes. The left-hand images show cancer cells labeled with quantum dots, and the right-hand drawings suggest how wavelength-resolved spectroscopy or spectral imaging could quantify surface and intracellular biomarkers

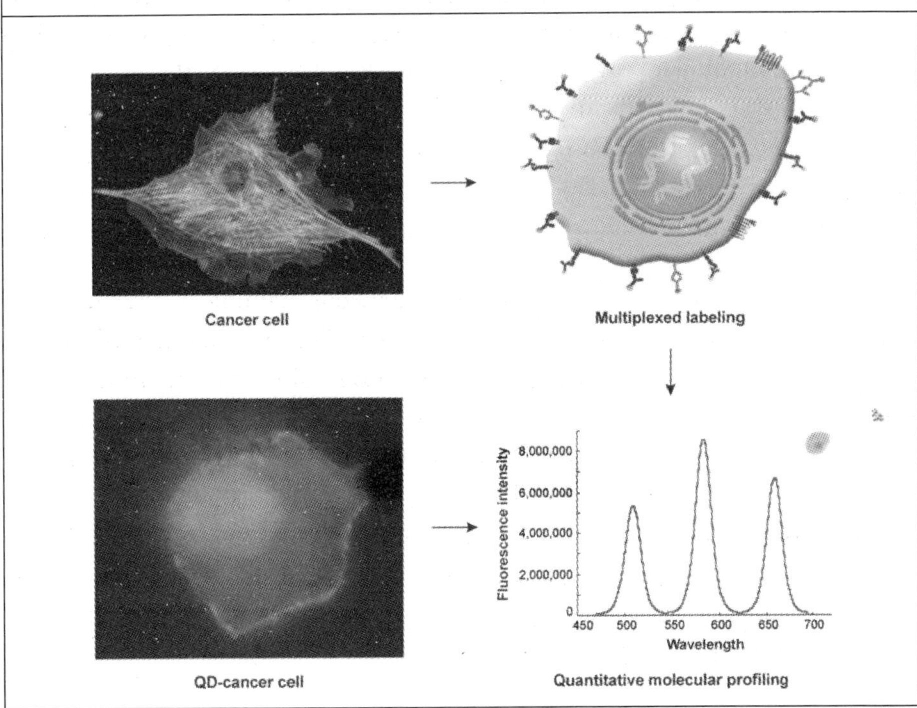

Correlation of Biomarkers with Cancer Behavior

Most studies on QD fluorescent labeling have been carried out with cells (both live and fixed)[79-81] or freshly harvested tissues[72,82]. However, the majority of available clinical specimens are archived, Formalin-Fixed Paraffin-Embedded (FFPE) tissues that might be several decades old. Because the clinical outcomes of these tissues are already known, it is of great value to use these specimens for examining the relationship between molecular profile and clinical outcome. Compared with cells or animal tissues, archived human specimens need special treatment, such as antigen retrieval, and their background autofluorescence is generally stronger. Our group has developed highly successful procedures for QD staining of archival FFPE tissue specimens. One example is to study the epithelial-mesenchymal transition (EMT) process in the progression of prostate cancer to the bone. EMT is a normal biological mechanism first reported in embryonic development and later found to be involved in cancer metastasis[83]. During EMT, cancer cells undergo phenotypical changes and become more invasive, characterized by changes in cellular adhesion molecules, particularly, an increase of N-cadherin and a loss of E-cadherin. Other important markers include the cytoskeleton proteins vimentin, cytokeratin 18, and RANKL. We have used QD-conjugated secondary antibodies for molecular profiling of two FFPE androgen-repressed prostate cancer cell lines ($ARCaP_e$ and $ARCaP_m$). These two cell lines represent two phenotypes at the two ends of the EMT process during prostate cancer progression. The $ARCaP_E$ is more epithelial-like and less invasive, whereas the $ARCaP_M$ has more mesenchymal characteristics and is more invasive[84].

QD staining studies have achieved simultaneous staining of four different biomarkers with expression profiles consistent with Western blot data (Figure 5). Moreover, QD staining provides spatial localization information (both inter- and intracellular), which is not possible with Western blot or other molecular biology techniques. We have also found that staining of FFPE cells requires longer incubation time (overnight at 4 C versus 1 h at room temperature) and a higher QD-secondary antibody concentration than that required for freshly fixed cells. Detailed methods and materials for QD bioconjugation, multiplexed tissue staining, and quantitative data analysis are provided in *Nature Protocols* (Nie and coworkers, 2(4): 1–15, 2007).

Figure 5: Multiplexed QD profiling of four tumor biomarkers using two FFPE prostate cancer cell lines (ARCaPe and ARCaPm) with distinct bone-metastasis behaviors. The four markers, all associated with epithelial-mesenchymal transition (EMT), are N-cadherin, EF (elongation factor)-1alpha, E-cadherin, and vimentin, and their corresponding QD colors are 565 nm, 605 nm, 655 nm, and 705 nm, respectively. The cell nuclei were counterstained blue by DAPI, and the spectra were captured under blue excitation

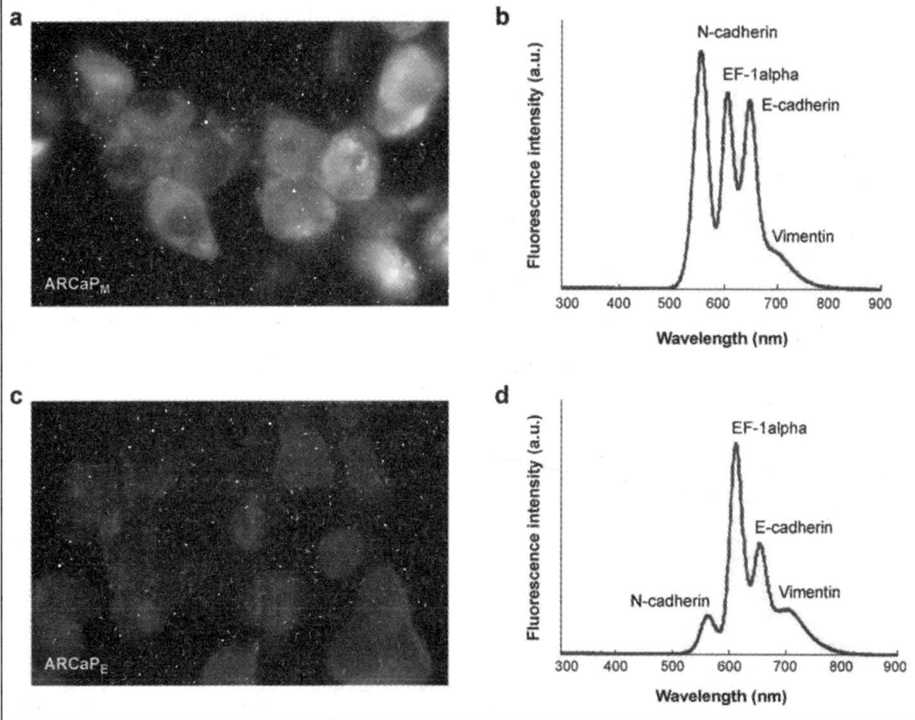

Note: (a) Color fluorescence image of highly metastatic prostate cancer cells (clone ARCaPm); (b) single-cell fluorescence spectrum obtained from image (a); (c) color fluorescence image of benign prostate cancer cells (clone ARCaPe); (d) single-cell spectrum obtained from image (c). The relative abundance of these markers is consistent with previous Western blot data. Note that individual cancer cells have heterogeneous expression patterns, and that the single-cell data in (b) and (d) are representative of a heterogeneous cell population.

For molecular profiling of clinical FFPE prostate specimens, we have selected four tumor antigens (mdm-2, p53, EGR-1, and p21) as a model system for technology development. These markers are known to be important in prostate cancer diagnosis and are correlated with tumor behavior[85,86]. As shown in Figure 6,

Figure 6: Multiplexed QD staining of archived FFPE clinical specimen from human prostate cancer patients and comparison between two different glands on the same tissue specimen. Four tumor biomarkers (mdm-2, p53, EGR-1, and p21) were labeled with four colors of QDs emitting at 565 nm, 605 nm, 655 nm, and 705 nm, respectively

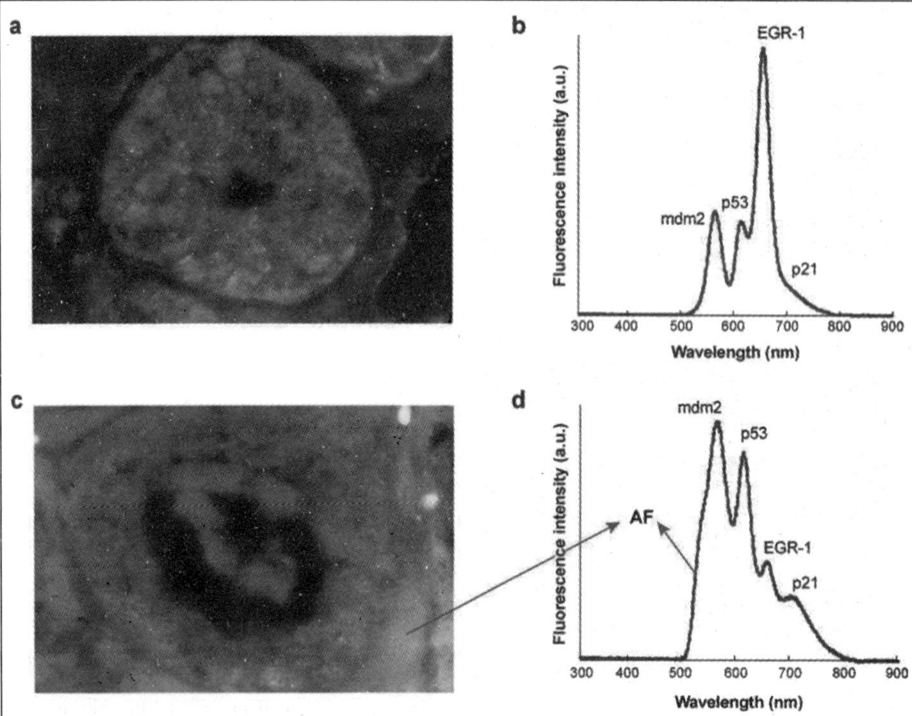

Note: (a) Color fluorescence image of QD-stained tissue specimens showing one prostate gland; (b) representative fluorescence spectrum obtained from individual cells in the gland (image a); (c) color fluorescence image of the same QD-stained tissue specimens showing a different gland; (d) representative fluorescence spectrum obtained from single cells in the second gland (image c). Note the distinct biomarker profiles for these two prostate glands, demonstrating the ability to resolve cellular populations in highly heterogeneous human tissue specimens. AF stands for autofluorescence and provides information on tissue morphology.

all four markers are detected in the tissue specimens, but the autofluorescence is higher than that observed in FFPE cells. In comparison with FFPE cells, clinical tissue specimens may require harsher antigen retrieval conditions (EDTA buffer versus citrate buffer) and generally have stronger autofluorescence. On the other hand, autofluorescence can be desirable by serving as a counterstain of tissue morphology. Autofluorescence can be separated from the QD signal by intentionally illuminating the sample to bleach it out while leaving the QDs bright enough for imaging and spectral analysis. In addition, spectral unmixing algorithms can be developed in-house or obtained commercially[87] for separating background fluorescence from true QD signals. These results demonstrate the feasibility of using QDs as fluorescent labels for molecular profiling of FFPE clinical specimens. With continuous efforts in optimizing the experimental conditions, we believe that QD probes hold great promise in multiplexed molecular profiling of clinical tissue specimens and for correlation studies of biomarkers and cancer behavior.

Early Cancer Detection

Bioconjugated particles and devices are also under development for early cancer detection in body fluids such as blood and serum. These nanoscale devices operate on the principles of selectively capturing cancer cells or target proteins. The sensors are often coated with a cancer-specific antibody or other biorecognition ligands so that the capture of a cancer cell or target protein yields an electrical, mechanical, or optical signal for detection. For example, microelectrical mechanical systems (MEMS) sensors rely on the deflection of nanometer-scale cantilever beams such as carbon nanotubes and metallic oxide nanobelts, structures that are sensitive to piconewton mechnical forces. Another promising area of research is the use of nanoparticles for detection and analysis of circulating tumor cells and biomarkers in blood/serum samples[88]. Vessella and coworkers[89] have demonstrated the ability to enrich for circulating cancer cells from both bone marrow aspirates and peripheral blood samples. However, the current systems are limited by the selectivity and efficiency to concentrate the rare cells for molecular assays. This is especially true for circulating cancer cells, which while present, are rarely isolated in quantities of greater than 1-2 cells per milliliter of blood. Through the combinatorial use of magnetic nanoparticles and semiconductor QDs, it is possible to increase the ability to capture and evaluate these rare circulating cancer cells. The resulting reporter signals would allow for the characterization of individual cancer cells for features associated with an aggressive phenotype (e.g., metastatic potential). The application of multiplexed nanoparticle probes would also allow for the interrogation of these cells for features related to treatment response.

Nanobarcodes

Mirkin and coworkers[90,91] reported an innovative approach for both protein and nucleic acid detection based on biobarcode-amplification (BCA). This approach uses both colloidal gold nanoparticles and magnetic microbeads, gold nanoparticles modified with both target capture strands and bar code strands that are subsequently hybridized to bar code DNA, and magnetic microparticles modified with target capture strands. In the presence of target DNA, the gold nanoparticles and the magnetic microbeads form sandwich structures that are magnetically separated from solution and are further washed to remove the unhybridized bar code DNA. The bar codes (hundreds to thousands per target) are detected by using a colorimetric method. This integrated capture and detection technology is—four to six orders of magnitude more sensitive than standard ELISA (enzyme-linked immunosorbent assay) for proteins and offers comparable sensitvities as PCR (polymerase chain reactions) for level nucleic acid targets[91].

Nanowires

Nanowires are available in metallic, semiconductor, magnetic, oxide, and polymer compositions and are promising as ultrasmall chemical and biological sensors[92,93]. Functionalized nanowires are coated with capture ligands such as antibodies or oligonucleotides. In the presence of target molecules, the specific binding between target molecule and capture molecule generates an immediate conductivity change within the nanowire that can be measured. Hahm *et al.,*[94] used silicon nanowire for ultrasensitive and selective detection of DNA. The surface of this nanowire device was coated with peptide nucleic acid (PNA) ligands for recognizing a mutation site in the cystic fibrosis transmembrane receptor gene. The achieved detection limit is on the order of 10 femtomolar (10×10^{-15} M). The same group has also developed nanowire arrays for multiplexed cancer biomarker detection[95], which consist of many individual nanowires each coated with a distinct surface receptor. These nanowire arrays allow simultaneous incorporation of control nanowires, which enables discrimination against false positives; they are also capable of selective and sensitive multiplexed detection of cancer biomarkers such as PSA, PSA-α1-antichymotrypsin, carcinoem-bryonic antigen, and mucin-1 in undiluted serum samples[95].

Carbon Nanotubes

Another type of nanodevice for biomarker detection is the Carbon Nanotube (CNT)[96]. Using single-walled carbon nanotubes as high-resolution atomic force microscopy (AFM) tips, Woolley *et al.,*[97] showed that specific sequences of kilobase-size DNA can be selectively detected from single-base mismatch sequences. Specifically, target DNA fragments were first hybridized with labeled (for instance, strepavidin-labeled) oligonucleotides, and then AFM was used to directly detect the presence and special location of the labels. This technique enabled the simple and direct detection of specific haplotypes that code for genetic disorders such as cancer. CNT-modified electrodes can amplify the electrochemical signal of guanuine bases, which has been used by Wang *et al.,*[98] for label-free electrochemical detection of DNA at nanomolar concentrations. More recent work has utilized CNTs coated with alkaline phosphatase as labels for amplified DNA and protein detection[99]. In this assay, CNTs play a dual amplification role in the recognition and transduction events, namely as carriers for numerous enzyme tags and for accumulating products of the enzymatic reaction.

Targeted Cancer Therapy

As noted above, most current anticancer agents do not greatly differentiate between cancerous and normal cells, leading to systemic toxicity and adverse effects. Consequently, systemic applications of these drugs often cause severe side effects in other tissues (such as bone marrow suppression, cardiomyopathy, and neurotoxicity), which greatly limits the maximal allowable dose of the drug. In addition, rapid elimination and widespread distribution into nontargeted organs and tissues requires the administration of a drug in large quantities, which is not economical and often complicated owing to nonspecific toxicity. Nanotechnology offers a more targeted approach and could thus provide significant benefits to cancer patients. In fact, the use of nanoparticles for drug delivery and targeting is likely one of the most exciting and clinically important applications of cancer nanotechnology. In this section, we discuss different targeting strategies for nanoscale drug delivery systems.

Passive Targeting

Rapid vascularization in fast-growing cancerous tissues is known to result in leaky, defective architecture and impaired lymphatic drainage. This structure allows an EPR effect[63,64,66,73], resulting in the accumulation of nanoparticles at the tumor site (Figure 7). For such a passive targeting mechanism to work, the size and surface properties of drug delivery nanoparticles must be controlled to avoid uptake by the reticuloendothelial system (RES)[100]. To maximize circulation times and targeting ability, the optimal size should be less than 100 nm in diameter and the surface should be hydrophilic to circumvent clearance by macrophages.

Figure 7: Schematic diagrams showing enhanced permeability and retention of nanoparticles in tumors. Normal tissue vasculatures are lined by tight endothelial cells, thereby preventing nanoparticle drugs from escaping or extravasation, whereas tumor tissue vasculatures are leaking and hyperpermeable allowing preferential accumulation of nanoparticles in the tumor interstitial space (called passive nanoparticle tumor targeting).

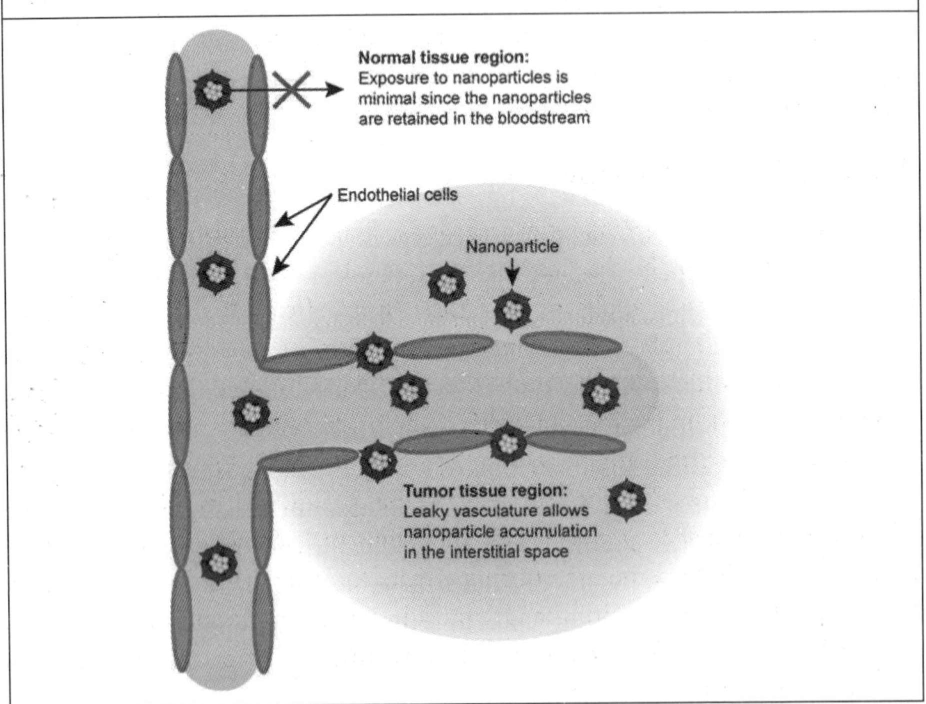

A hydrophilic surface of the nanoparticles safeguards against plasma protein adsorption and can be achieved through hydrophilic polymer coatings such as PEG, poloxamines, poloxamers, polysaccharides, or through the use of branched or block amphiphilic copolymers[101-104]. The covalent linkage of amphiphilic copolymers (polylactic acid, polycaprolactone, polycyanonacrylate chemically coupled to PEG) is generally preferred, as it avoids aggregation and ligand desorption when in contact with blood components.

An alternative passive targeting strategy is to utilize the unique tumor environment in a scheme called tumor-activated prodrug therapy. The drug is conjugated to a tumor-specific molecule and remains inactive until it reaches the target[105]. Over-expression of the matrix metalloproteinase (MMP) MMP-2 in melanoma has been shown in a number of preclinical as well as clinical investigations. Mansour *et al.,*[106] reported a water-soluble maleimide derivative of doxorubicin (DOX) incorporating an MMP-2-specific peptide sequence (Gly-Pro-Leu-Gly-Ile-Ala-Gly-Gln) that rapidly and selectively binds to the cysteine-34 position of circulating albumin. The albumin-DOX conjugate is efficiently and specifically cleaved by MMP-2, releasing a DOX tetrapeptide (Ile-Ala-Gly-Gln-DOX) and subsequently DOX. pH and redox potential have been also explored as drug release triggers at the tumor site[107]. Another passive targeting method is the direct local delivery of anticancer agents to tumors. This approach has the obvious advantage of excluding the drug from the systemic circulation. However, administration can be highly invasive, as it involves injections or surgical procedures. For some tumors, such as lung cancers, that are difficult to access, the technique is nearly impossible to use.

Active Targeting

Active targeting is usually achieved by conjugating to the nanoparticle a targeting component that provides preferential accumulation of nanoparticles in the tumor-bearing organ, in the tumor itself, individual cancer cells, or intracellular organelles inside cancer cells. This approach is based on specific interactions, such as lectin-carbohydrate, ligand-receptor, and antibody-antigen[108]. Lectin-carbohydrate is one of the classic examples of targeted drug delivery[109]. Lectins are proteins of nonimmunological origin, capable of recognizing and binding to glycoproteins expressed on cell surfaces. Lectin interactions with certain carbohydrates are very specific. Carbohydrate moieties can be used to target drug delivery systems to lectins (direct lectin targeting), and lectins can be used as targeting moieties to target cell surface carbohydrates (reverse lectin targeting). However, drug delivery systems based on lectin-carbohydrate have mainly been developed to target whole organs[110], which can pose harm to normal cells. Therefore, in most cases the targeting moiety is directed toward specific receptors or antigens expressed on the plasma membrane or elsewhere at the tumor site.

The overexpression of receptors or antigens in many human cancers lends itself to efficient drug uptake via receptor-mediated endocytosis (Figure 8). Because

Figure 8: Nanoparticle drug delivery and targeting using receptor-mediated endocytosis. The nanoparticle drug is internalized by tumor cells through ligand-receptor interaction. Depending on the design of the cleavable bond, the drug will be released intracellularly on exposure to lysosomal enzymes or lower pH

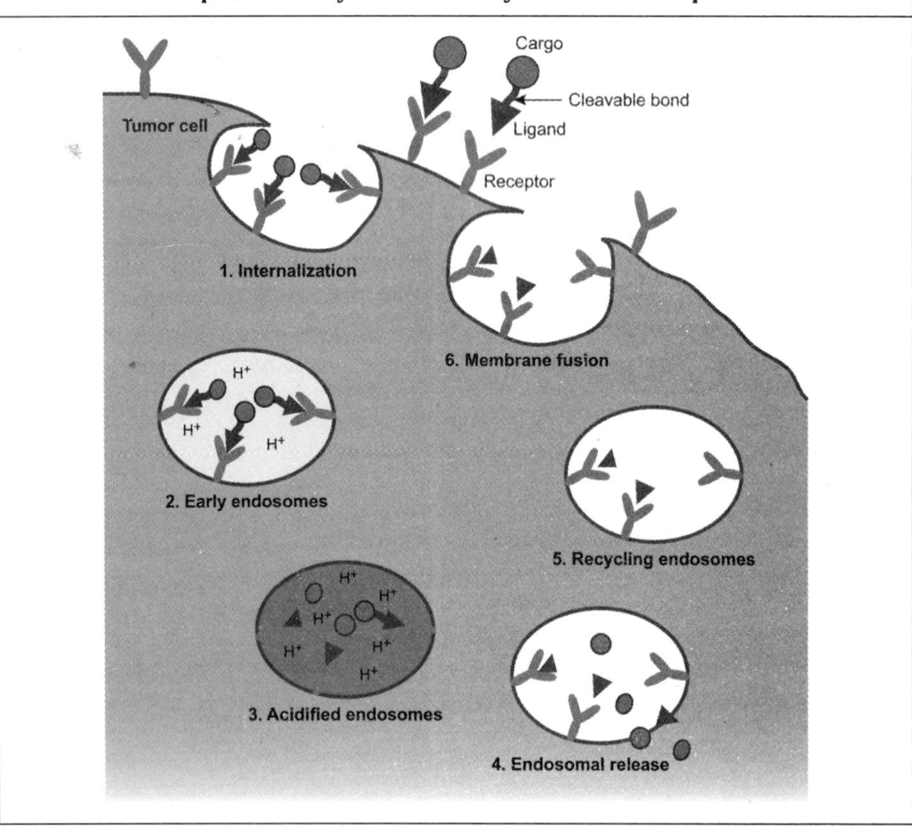

gly-coproteins cannot remove polymer-drug conjugates that have entered the cells via endocytosis[111,112], this active targeting mechanism provides an alternative route for overcoming multiple drug resistance (MDR)[113-118]

The cell surface receptor for folate is inaccessible from the circulation to healthy cells owing to its location on the apical membrane of polarized epithelia, but it is over-expressed on the surface of various cancers, including ovary, brain, kidney, breast, and lung malignancies[119,120]. Surface plasmon resonance studies revealed that folate-conjugated PEGylated cyanoacrylate nanoparticles had a tenfold higher affinity for the folate receptor than free folate did[121]. Folate receptors are often organized in clusters and bind preferably to the multivalent forms of the ligand. Furthermore, con-focal microscopy demonstrated selective uptake and endocytosis of folate-conjugated nanoparticles by tumor cells bearing folate receptors. Interest in exploiting folate receptor targeting in cancer therapy and diagnosis has rapidly increased, as attested by many conjugated systems, including proteins, liposomes, imaging agents, and neutron activation compounds[119,120].

Nanoparticle Drugs

Nanotechnology is beginning to change the scale and methods of drug delivery (Figure 9). Therapeutic and diagnostic agents can be encapsulated, covalently attached, or adsorbed onto nanoparticles. These approaches can easily overcome drug solubility issues, which has significant implications because more than 40% of active substances being identified through combinatorial screening programs are poorly soluble in water[122]. Conventional and most current formulations of such drugs are frequently plagued with problems such as poor and inconsistent bioavailability. The widely used attempt at enhancing solubility is to generate a salt. For nonionizable compounds, micronization, soft-gel technology, cosolvents, surfactants, or complexing agents have been used[123]. Because it is faster and more cost effective to reformulate the drug than to develop a new one, a broadly based technology applicable to poorly water-soluble drugs could make a tremendous impact.

For decades, researchers have been developing new anticancer agents and new formulations for delivering chemotherapy drugs[112]. Paclitaxel (Taxol™) is one of

Figure 9: Illustration showing self-assembled polymeric nanoparticles with dual tumor-targeting and therapeutic functions (upper panel) and delivery of the nanoparticle drugs by receptor-mediated endocytosis and controlled drug release inside the cytoplasm (lower panel)

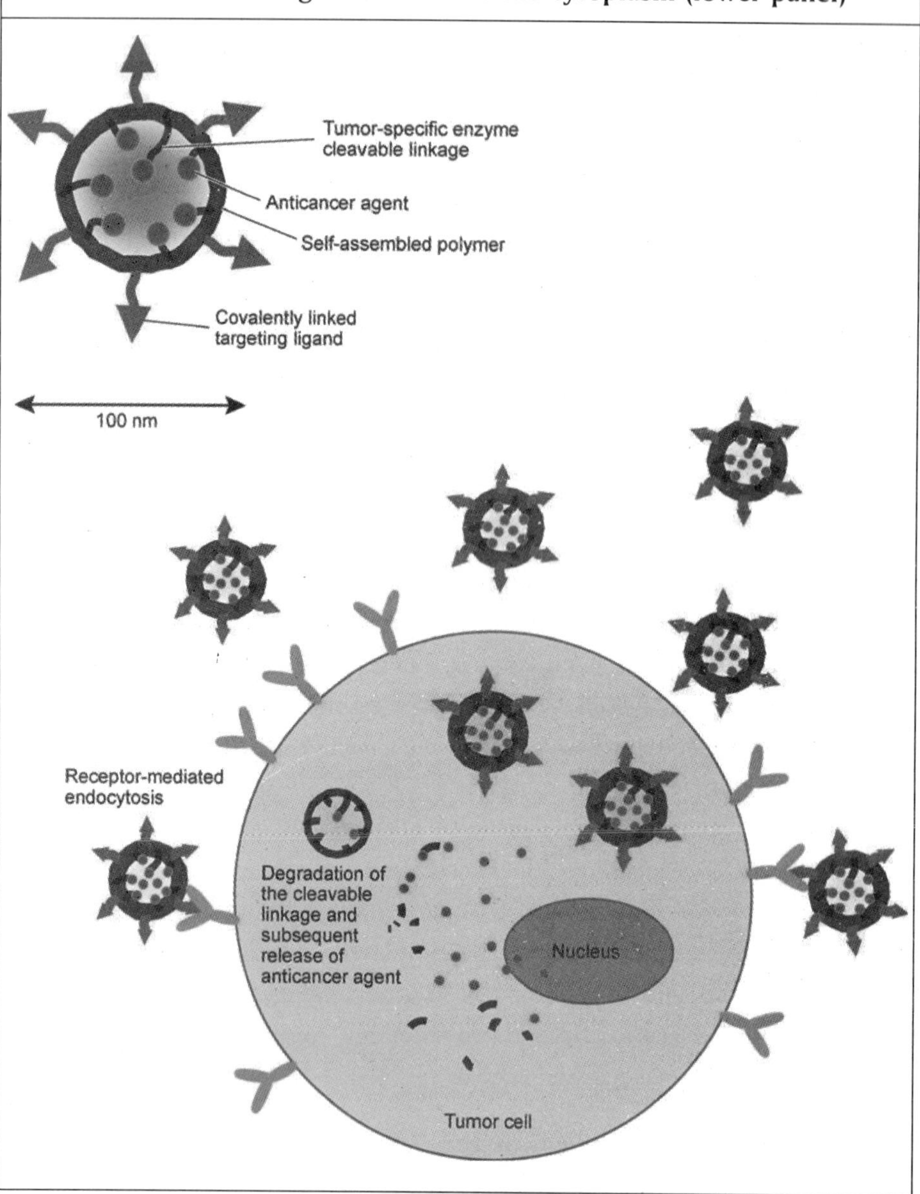

the most widely used anticancer drugs in the clinic. It is a microtubule-stabilizing agent that promotes tubulin polymerization, disrupting cell division and leading to cell death[(124,125)]. It displays neoplastic activity against primary epithelial ovarian carcinoma and breast, colon, and lung cancers. Because it is poorly soluble in aqueous solution, the formulation available currently is Chremophor EL (polyethoxylated castor oil) and ethanol[(126)]. In a new formulation approach used in Abraxane™, recently approved by the FDA to treat metastatic breast cancer, paclitaxel was conjugated to albumin nanoparticles[(127,128)]. The formulation is very effective in circumventing side effects of the highly toxic Chremophor EL, which include hypersensitivity reactions, nephrotoxicity, and neurotoxicity[(126,129)]. Although the SPACR (secreted protein, acidic, cysteine-rich, also called osteonectin) protein is believed to improve albumin drug uptake, this nanoparticulate drug still exhibits significant side effects (see FDA-Approved Nanoparticle Drug—Abraxane).

Fda-Approved Nanoparticle Drug—Abraxane

The Food and Drug Administration (FDA) recently approved Abraxane™, an albumin-paclitaxel (Taxol™) nanoparticle for the treatment of metastatic breast cancer. A Phase I clinical trial determined that the maximum tolerated dose (MTD) of single-agent albumin-bound paclitaxel every 3 weeks was 300 mg/m^2 in patients with solid tumors (breast cancer and melanoma). A second Phase I trial, reported at the 2004 ASCO Annual meeting, demonstrated 5 responses among 39 pretreated patients with advanced solid tumors, including 1 response in a patient with NSCLC, 3 responses in patients with ovarian cancer, and 1 in breast cancer. The dose-limiting toxicity was myelosuppression, the MTD was 270 mg/m^2, and premedication was not required. Subsequent use of Abraxane in both Phase II and Phase III trails proved that this new formulation was far superior to Taxol™. In a randomized, open-labeled trial of 454 patients with metastatic breast cancer, the overall response rate for ABI-007 was 33%, compared with 19% for Taxol™. Median time to progression was 21.9 weeks for ABI-007, versus 16.1 weeks for Taxol™. Overall side effects were fewer ABI-007, even though it delivered a 50% higher dose of the active agent Taxol™.

For enhanced tumor-specific targeting, the differences between cancerous cells and normal cells may be exploited. By virtue of their small size, nanoparticles entail a high surface area that not only paves the way for more efficient drug release but also a better strategy for functionalization. There is a growing body of knowledge of unique cancer markers thanks to recent advances in proteomics and genomics. They form the basis of complex interactions between bioconjugated nanoparticles and cancer cells. Carrier design and targeting strategies may vary according to the type, developmental stage, and location of cancer[(130)]. There is much synergy between imaging and nanotechnology in biomedical applications. Many of the principles used to target delivery of drugs to cancer may also be applied to target imaging and diagnostic agents to enhance detection sensitivity in medical imaging. With engineered multifunctional nanoparticles (Figures 10), the full in vivo potential of cancer nanotechnology in targeted drug delivery and imaging can be realized.

Figure 10: Multifunctional nanoparticles for integrated cancer imaging and therapy. A truly exciting feature of cancer nanotechnology is that drug delivery, treatment efficacy, and toxicity could be monitored by using embedded imaging agents

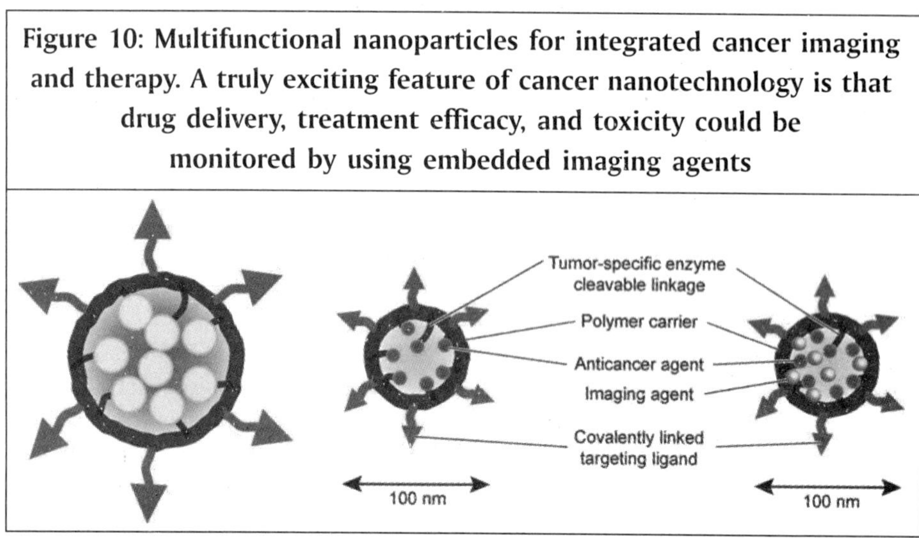

Future Directions

Nanotechnology has become an enabling technology for personalized oncology in which cancer detection, diagnosis, and therapy are tailored to each mor molecular profile, and also for predictive oncology in which genetic/molecular markers are used to predict disease development, progression, and clinical outcomes. In recognition of its potential impact in cancer research, the US National Cancer Institute (NCI) has recently funded eight national Centers of Cancer Nanotechnol-ogy Excellence (CCNE). Looking into the future, there are a number of research themes or directions that are particularly promising but require concerted effort for success. The first direction is the design and development of nanoparticles with monofunctions, dual functions, three functions, or multiple functions. For cancer and other medical applications, important functions include imaging (single or dual-modality), therapy (single drug or combination of two or more drugs), and targeting (one or more ligands). With each added function, nanoparticles could be designed to have novel properties and applications. For example, binary nanoparticles with two functions could be developed for molecular imaging, targeted therapy, or for simultaneous imaging and therapy (but without targeting). Bioconjugated QDs with both targeting and imaging functions will be used for targeted tumor imaging and molecular profiling applications. Conversely, ternary nanoparticles with three functions could be designed for simultaneous imaging and therapy with targeting, targeted dual-modality imaging, or targeted dual-drug therapy. Quaternary nanoparticles with four functions can be conceptualized in the future to have the abilities of tumor targeting, dual-drug therapy, and imaging. The second direction is nanoparticle molecular profiling (nanotyping) for clinical oncology; that is, the use of bioconjugated nanoparticle probes to predict cancer behavior, clinical outcome, and treatment response and to individualize therapy. This should start with retrospective studies of archived specimens because the patient outcome is already known for these specimens. The key hypotheses to be tested are that nanotyping a panel of tumor markers will allow more accurate correlations than single tumor markers, and that the combination of nanotyping tumor gene expression and host stroma are both important in defining the aggressive phenotypes of cancer as well as determining the response of early stage disease to treatment (chemotherapy, radiation, or surgery). The third important direction is to study nanoparticle distribution, excretion, metabolism, and pharmacodynamics in in vivo animal models. These investigations will be very important in the development of nanoparticles for clinical applications in cancer imaging or therapy.

Summary Points

1. Nanometer-sized particles have novel optical, electronic, magnetic, or structural properties and are currently under intense development for applications in cancer, cardiovascular diseases, and degenerative neurological disorders such as Alzheimer's disease.
2. Quantum dots are just one type of nanoparticle with novel optical properties such as size-tunable emission, improved signal brightness, resistance against photobleaching, and simultaneous excitation of multiple fluorescence colors.
3. Dual-modality and multifunctional probes are being developed by attaching molecular moieties with imaging, therapeutic, and targeting functions to nanostructured scaffolds. These integrated nanoparticle probes may allow simultaneous imaging and therapy of tumors and cardiovascular plaques in live animal models.
4. Antibody-conjugated multicolor quantum dots have been used for multiplexed molecular profiling of cancer cells and clinical tissue specimens, and for correlation of a panel of 4 – 5 biomarkers with cancer behavior and patient outcome.
5. Bionanobarcodes, nanocantilevers, and nanowires are promising technologies for early cancer detection and screening in blood and serum samples.
6. Targeted nanoparticle drugs offer significant advantages in improving cancer therapeutic efficacy and simultaneously reducing drug toxicity.
7. Dual-and multimodality nanoparticles are being developed by attaching molecular moieties with imaging, therapeutic, and targeting functions to nanometer-scaled scaffolds for simultaneous imaging and therapy of tumors.
8. Future work needs to address the potential long-term toxicity, degradation, and metabolism of nanoparticle agents, to identify and develop new biomarker-probe systems, and to develop multifunctional nanoscale platforms for integrated imaging, detection, and therapy.

Disclosure Statement

The authors are not aware of any biases that might be perceived as affectin the objectivity of this review.

Acknowledgments

We are grateful to Mr. Amit Agrawal, Dr. Gang Ruan, Dr. Leland Chung, Dr. Ruth O'Regan, and Dr. Dong Shin for May D. Wang (Core Director, Emory-Georgia Tech Nanotechnology Center for Personalized and Predictive Oncology) for help on cancer informatics and biocom-puting. Our research was supported by NIH grants (P20 GM072069, R01 CA108468-01, and U54CA119338) and the Georgia Cancer Coalition Distinguished Cancer Scholars Program (to SN).

(Shuming Nie, Yun Xing, Gloria J Kim, and JonathanW Simons are associated with Department of Biomedical Engineering and theWinship Cancer Institute, Emory University and Georgia Institute of Technology, Atlanta, Georgia.)

Literature Cited

1 Hanahan D, Weinberg RA. 2000. "The hallmarks of cancer". *Cell* 100:57-70.

2 Hahn WC, Weinberg RA. 2002. "Modelling the molecular circuitry of cancer". *Nat. Rev. Cancer* 2:331-41.

3 Liotta L, Petricoin E. 2000. "Molecular profiling of human cancer". *Nat. Rev. Genet.* 1:48-56.

4 Petricoin EF, Zoon KC, Kohn EC, Barrett JC, Liotta LA. 2002. "Clinical pro-teomics: translating benchside promise into bedside reality". *Nat. Rev. Drug Dis-cov.* 1:683-95.

5 Menon U, Jacobs IJ. 2000. "Recent developments in ovarian cancer screening". *Curr. Opin. Obstet. Gynecol.* 12:39-42.

6 Ferrari M. 2005. "Cancer nanotechnology: opportunities and challenges". *Nat. Rev. Cancer* 5:161-71.

7 Srinivas PR, Barker P, Srivastava S. 2002. "Nanotechnology in early detection of cancer". *Lab. Invest.* 82:657-62.

8 Henglein A. 1989. "Small-particle research—physicochemical properties of extremely small colloidal metal and semiconductor particles". *Chem. Rev.* 89:1861-73

9 Schmid G. 1992. "Large clusters and colloids—metals in the embryonic state". *Chem. Rev.* 92:1709-27.

10 Niemeyer CM. 2001. "Nanoparticles, proteins, and nucleic acids: biotechnology meets materials science". *Angew. Chem. Int. Ed. Engl.* 40:4128-58.

11 Alivisatos P. 2004. The use of nanocrystals in biological detection. *Nat. Biotechnol.* 22:47-52.

12 Alivisatos AP. 1996. "Semiconductor clusters, nanocrystals, and quantum dots". *Science* 271:933-37.

13 Alivisatos AP, Gu WW, Larabell C. 2005. "Quantum dots as cellular probes". *Annu. Rev. Biomed. Eng.* 7:55-76.

14 Pinaud F, Michalet X, Bentolila LA, Tsay JM, Doose S, *et al.*, 2006. "Advances in fluorescence imaging with quantum dot bio-probes". *Biomaterials* 27:1679-87.

15 Michalet X, Pinaud FF, Bentolila LA, Tsay JM, Doose S, *et al.*, 2005. "Quantum dots for live cells, in vivo imaging, and diagnostics". *Science* 307:538-44.

16 Gao XH, Yang LL, Petros JA, Marshal FF, Simons JW, Nie SM. 2005. "In vivo molecular and cellular imaging with quantum dots". *Curr. Opin. Biotechnol.* 16:63-72.

17 Smith AM, Gao X, Nie S. 2004. "Quantum dot nanocrystals for in vivo molecular and cellular imaging". *Photochem. Photobiol.* 80:377-85.

18 Chan WCW, Maxwell DJ, Gao XH, Bailey RE, Han MY, Nie SM. 2002. "Luminescent quantum dots for multiplexed biological detection and imaging". *Curr. Opin. Biotechnol.* 13:40-46.

19 Harisinghani MG, Barentsz J, Hahn PF, Deserno WM, Tabatabaei S, *et al.*, 2003. "Noninvasive detection of clinically occult lymph-node metastases in prostate cancer". *N. Engl. J. Med.* 348:2491-99.

20 Hood JD, Bednarski M, Frausto R, Guccione S, Reisfeld RA, *et al.*, 2002. "Tumor regression by targeted gene delivery to the neovasculature". *Science* 296:2404-7.

21 Walt DR. 2002. "Imaging optical sensor arrays". *Curr. Opin. Chem. Biol.* 6:689-95.

22 Nicewarner-Pena SR, Freeman RG, Reiss BD, He L, Pena DJ, *et al.*, 2001. Submicrometer metallic barcodes. *Science* 294:137-41.

23 Cunin F, Schmedake TA, Link JR, Li YY, Koh J, *et al.*, 2002. "Biomolecular screening with encoded porous-silicon photonic crystals". *Nat. Mater.* 1:39-41.

24 Dejneka MJ, Streltsov A, Pal S, Frutos AG, Powell CL, *et al.*, 2003. "Rare earth-doped glass microbarcodes". *Proc. Natl. Acad. Sci. USA* 100:389-93.

25 Cao YWC, Jin RC, Mirkin CA. 2002. "Nanoparticles with Raman spectroscopic fingerprints for DNA and RNA detection". *Science* 297:1536-40.

26 Golub TR, Slonim DK, Tamayo P, Huard C, Gaasenbeek M, *et al.*, 1999. "Molecular classification of cancer: class discovery and class prediction by gene expression monitoring". *Science* 286:531-37.

27 Ross DT, Scherf U, Eisen MB, Perou CM, Rees C, *et al.*, 2000. "Systematic variation in gene expression patterns in human cancer cell lines". *Nat. Genet.* 24:227-35.

28 Alizadeh AA, Eisen MB, Davis RE, Ma C, Lossos IS, *et al.*, 2000. "Distinct types of diffuse large B-cell lymphoma identified by gene expression profiling". *Nature* 403:503-11.

29 Perou CM, Sorlie T, Eisen MB, van de Rijn M, Jeffrey SS, *et al.*, 2000. "Molecular portraits of human breast tumours". *Nature* 406:747-52.

30 Bittner M, Meitzer P, Chen Y, Jiang Y, Seftor E, *et al.*, 2000. "Molecular classifi-cation of cutaneous malignant melanoma by gene expression profiling". *Nature* 406:536-40.

31 Dhanasekaran SM, Barrette TR, Ghosh D, Shah R, Varambally S, *et al.*, 2001. "Delineation of prognostic biomarkers in prostate cancer". *Nature* 412:822-26.

32 Kuefer R, Varambally S, Zhou M, Lucas PC, Loeffler M, *et al.*, 2002. □α-Methyl acyl-CoA racemase: expression levels of this novel cancer biomarker depend on tumor differentiation". *Am. J. Pathol.* 161:841-48.

33 Rhodes DR, Barrette TR, Rubin MA, Ghosh D, Chinnaiyan AM. 2002. "Meta-analysis of microarrays: interstudy validation of gene expression profiles reveals pathway dysregulation in prostate cancer". *Cancer Res.* 62:4427-33.

34 Rhodes DR, Sanda MG, Otte AP, Chinnaiyan AM, Rubin MA. 2003. "Multiplex biomarker approach for determining risk of prostate-specific antigen-defined recurrence of prostate cancer". *J. Natl. Cancer Inst.* 95:661-68.

35 Rubin MA. 2001. "Use of laser capture microdissection, cDNA microarrays, and tissue microarrays in advancing our understanding of prostate cancer". *J. Pathol.* 195:80-86.

36 Rubin MA, Dunn R, Strawderman M, Pienta K. 2002. "Tissue microarray sampling strategy for prostate cancer biomarker analysis". *Am. J. Surg. Pathol.* 26(3):312-19.

37 Rubin MA, Zhou M, Dhanasekaran SM, Varambally S, Barrette TR, *et al.*, 2002. "α-Methylacyl coenzyme A racemase as a tissue biomarker for prostate cancer". *JAMA* 287:1662-70.

38 Varambally S, Dhanasekaran SM, Zhou M, Barrette TR, Kumar-Sinha C, *et al.*, 2002. "The polycomb group protein EZH2 is involved in progression of prostate cancer". *Nature* 419:624-29.

39 Nelson WG, De Marzo AM, Isaacs WB. 2003. "Mechanisms of disease: prostate cancer". *N Engl. J Med.* 349:366-81

40 Gonzalgo ML, Pavlovich CP, Lee SM, Nelson WG. 2003. "Prostate cancer detection by GSTP1 methylation analysis of postbiopsy urine specimens". *Clin. Cancer Res.* 9:2673-77.

41 Weinshilboum R, Wang LW. 2004. "Pharmacogenomics: bench to bedside". *Nat. Rev. Drug Discov.* 3:739-48.

42 Evans WE, Relling MV. 2004. "Moving towards individualized medicine with pharmacogenomics". *Nature* 429: 464-68.

43 Lynch TJ, Bell DW, Sordella R, Gurubhagavatula S, Okimoto RA, *et al.*, 2004. "Activating mutations in the epidermal growth factor receptor underlying responsiveness of nonsmall-cell lung cancer to gefitinib". *N. Engl. J. Med.* 350: 2129-39.

44 Paez JG, Janne PA, Lee JC, Tracy S, Greulich H, *et al.*, 2004. "EGFR mutations in lung cancer: correlation with clinical response to gefitinib therapy". *Science* 304:1497-500.

45 Michener CM, Ardekani AM, Petricoin EF, Liotta LA, Kohn EC. 2002. "Ge-nomics and proteomics: application of novel technology to early detection and prevention of cancer". *Cancer Detect. Prev.* 26: 249-55.

46 Srinivas PR, Verma M, Zhao Y, Srivastava S. 2002. "Proteomics for cancer biomarker discovery". *Clin. Chem.* 48:1160-69.

47 Yu WW, Qu LH, Guo WZ, Peng XG. 2003. "Experimental determination of the extinction coefficient of CdTe, CdSe, and CdS nanocrystals". *Chem. Mater.* 15: 2854-60.

48 Kim SW, Zimmer JP, Ohnishi S, Tracy JB, Frangioni JV, Bawendi MG. 2005. "Engineering InAsxP1-x/InP/ZnSe III-V alloyed core/shell quantum dots for the near-infrared". *J. Am. Chem. Soc.* 127:10526-32.

49 Gao XH, Cui YY, Levenson RM, Chung LWK, Nie SM. 2004. "In vivo cancer targeting and imaging with semiconductor quantum dots". *Nat. Biotechnol.* 22:969-76.

50 Ntziachristos V, Bremer C, Weissleder R. 2003. "Fluorescence imaging with near-infrared light: new technological advances that enable in vivo molecular imaging". *Eur. Radiol.* 13:195-208.

51 Ntziachristos V, Schellenberger EA, Ripoll J, Yessayan D, Graves E, *et al.*, 2004. "Visualization of antitumor treatment by means of fluorescence molecular tomography with an annexin V-Cy5.5 conjugate". *Proc. Natl. Acad. Sci. USA* 101:12294-99.

52 Kircher MF, Weissleder R, Josephson L. 2004. "A dual fluorochrome probe for imaging proteases. *Bioconjug". Chem.* 15:242-48.

53 Schellenberger EA, Sosnovik D, Weissleder R, Josephson L. 2004. "Magneto/ optical annexin V, a multimodal protein". *Bioconjug. Chem.* 15:1062-67.

54 Wang DS, He JB, Rosenzweig N, Rosenzweig Z. 2004. "Superparamagnetic Fe2O3 Beads-CdSe/ZnS quantum dots core-shell nanocomposite particles for cell separation". *Nano Lett.* 4:409-13.

55 Gu H, Zheng R, Zhang X, Xu B. 2004. "Facile one-pot synthesis of bifunc-tional heterodimers of nanoparticles: a conjugate of quantum dot and magnetic nanoparticles". *J. Am. Chem. Soc.* 126:5664-65.

56 Mulder WJM, Koole R, Brandwijk RJ, Storm G, Chin PTK, *et al.*, 2006. "Quantum dots with a paramagnetic coating as a bimodal molecular imaging probe". *Nano Lett.* 6:1-6

57 Van Tilborg GAF, Mulder WJM, Chin PTK, Storm G, Reutelingsperger CP, *et al.*, 2006. "Annexin A5-conjugated quantum dots with a paramagnetic lipidic coating for the multimodal detection of apoptotic cells". *Bioconjug. Chem.* 17:865-68.

58 Quintana A, Raczka E, Piehler L, Lee I, Myc A, *et al.*, 2002. "Design and function of a dendrimer-based therapeutic nanodevice targeted to tumor cells through the folate receptor". *Pharm. Res.* 19:1310-16.

59 Torchilin VP, Trubetskoy VS, Milshteyn AM, Canillo J, Wolf GL, *et al.*, 1994. "Targeted delivery of diagnostic agents by surface-modified liposomes". *J. Control. Release* 28:45-58.

60 Torchilin V, Babich J, Weissig V. 2000. "Liposomes and micelles to target the blood pool for imaging purposes". *J. Liposome Res.* 10:483-99.

61 Patri AK, Myc A, Beals J, Thomas TP, Bander NH, Baker JR. 2004. "Synthesis and in vitro testing of J591 antibody-dendrimer conjugates for targeted prostate cancer therapy". *Bioconjug. Chem.* 15:1174-81.

62 Buck SM, Koo YEL, Park E, Xu H, Philbert MA, *et al.*, 2004. "Optochemical nanosensor PEBBLEs: photonic explorers for bioanalysis with biologically localized embedding". *Curr. Opin. Chem. Biol.* 8:540-46.

63. Matsumura Y, Maeda H. 1986. "A new concept for macromolecular therapeutics in cancer chemotherapy: mechanism of tumoritropic accumulation of proteins and the antitumor agent smancs". *Cancer Res.* 46:6387-92.

64 Duncan R. 2003. "The dawning era of polymer therapeutics". *Nat. Rev. Drug Discov.* 2:347-60.

65 Jain RK. 1999. "Transport of molecules, particles, and cells in solid tumors". *Annu. Rev. Biomed. Eng.* 1:241-63.

66 Jain RK. 2001. "Delivery of molecular medicine to solid tumors: lessons from in vivo imaging of gene expression and function". *J. Control Release* 74:7-25.

67 Ballou B, Lagerholm BC, Ernst LA, Bruchez MP, Waggoner AS. 2004. "Noninvasive imaging of quantum dots in mice". *Bioconjug. Chem.* 15:79-86.

68 Larson DR, Zipfel WR, Williams RM, Clark SW, Bruchez MP, *et al.*, 2003. "Water-soluble quantum dots for multiphoton fluorescence imaging in vivo. *Science* 300:1434-36.

69 Stroh M, Zimmer JP, Duda DG, Levchenko TS, Cohen KS, *et al.*, 2005. "Quantum dots spectrally distinguish multiple species within the tumor milieu in vivo". *Nat. Med.* 11:678-82.

70 Kim S, Lim YT, Soltesz EG, De Grand AM, Lee J, *et al.*, 2004. "Near-infrared fluorescent type II quantum dots for sentinel lymph node mapping". *Nat. Biotechnol.* 22:93-97.

71 Akerman ME, Chan WCW, Laakkonen P, Bhatia SN, Ruoslahti E. 2002. "Nanocrystal targeting in vivo". *Proc. Natl. Acad. Sci. USA* 99:12617-21.

72 Ness JM, Akhtar RS, Latham CB, Roth KA. 2003. "Combined tyramide signal amplification and quantum dots for sensitive and photostable immunofluores-cence detection". *J. Histochem. Cytochem.* 51:981-87.

73 Jain RK. 1999. "Understanding barriers to drug delivery: high resolution in vivo imaging is key". *Clin. Cancer Res.* 5:1605-6.

74 Schulke N, Varlamova OA, Donovan GP, Ma DS, Gardner JP, *et al.*, 2003. "The homodimer of prostate-specific membrane antigen is a functional target for cancer therapy". *Proc. Natl. Acad. Sci. USA* 100:12590-95.

75 Bander NH, Trabulsi EJ, Kostakoglu L, Yao D, Vallabhajosula S, *et al.*, 2003. "Targeting metastatic prostate cancer with radiolabeled monoclonal antibody J591 to the extracellular domain of prostate specific membrane antigen". *J. Urol.* 170:1717-21.

76 Derfus AM, Chan WCW, Bhatia SN. 2004. "Probing the cytotoxicity of semiconductor quantum dots". *Nano Lett.* 4:11-18.

77 Ipe BI, Lehnig M, Niemeyer CM. 2005. "On the generation of free radical species from quantum dots". *Small* 1:706-9.

78 Kirchner C, Liedl T, Kudera S, Pellegrino T, Munoz Javier A, *et al.*, 2005. "Cytotoxicity of colloidal CdSe and CdSe/ZnS nanoparticles". *Nano Lett.* 5:331-38.

79 Lidke DS, Nagy P, Heintzmann R, Arndt-Jovin DJ, Post JN, *et al.*, 2004. "Quantum dot ligands provide new insights into erbB/HER receptor-mediated signal transduction". *Nat. Biotechnol.* 22:198-203.

80 Wu XY, Liu HJ, Liu JQ, Haley KN, Treadway JA, *et al.*, 2003. "Immunoflu-orescent labeling of cancer marker Her2 and other cellular targets with semiconductor quantum dots". *Nat. Biotechnol.* 21:41-46.

81 Tokumasu F, Dvorak J. 2003. "Development and application of quantum dots for immunocytochemistry of human erythrocytes". *J. Microsc.* 211:256-61.

82 Ferrara DE, Weiss D, Carnell PH, Vito RP, Vega D, *et al.*, 2006. "Quantitative 3D fluorescence technique for the analysis of en face preparations of arterial walls using quantum dot nanocrystals and two-photon excitation laser scanning microscopy". *Am. J. Physiol. Regul. Integr. Comp. Physiol.* 290:R114-23.

83 Huber MA, Kraut N, Beug H. 2005. "Molecular requirements for epithelial-mesenchymal transition during tumor progression". *Curr. Opin. Cell Biol.* 17:548-58.

84 Zhau HY, Chang SM, Chen BQ, Wang Y, Zhang H, *et al.*, 1996. "Androgen-repressed phenotype in human prostate cancer". *Proc. Natl. Acad. Sci. USA* 93:15152-57.

85 Hernandez I, Maddison LA, Wei YL, DeMayo F, Petras T, *et al.*, 2003. "Prostate-specific expression of p53 (R172L) differentially regulates p21, Bax, and mdm2 to inhibit prostate cancer progression and prolong survival". *Mol. Cancer Res.* 1:1036-47.

86 Mora GR, Olivier KR, Mitchell RFJ, Jenkins RB, Tindall DJ. 2005. "Regulation of expression of the early growth response gene-1 (EGR-1) in malignant and benign cells of the prostate". *Prostate* 63:198-207.

87 Mansfield JR, Gossage KW, Hoyt CC, Levenson RM. 2005. "Autofluorescence removal, multiplexing, and automated analysis methods for in-vivo fluorescence imaging". *J. Biomed. Opt.* 10(4):41207.

88 Cristofanilli M, Budd GT, Ellis MJ, Stopeck A, Matera J, *et al.*, 2004. "Circulating tumor cells, disease progression, and survival in metastatic breast cancer". *N. Engl. J. Med.* 351:781-91.

89 Ellis WJ, Pfitzenmaier J, Colli J, Arfman E, Lange PH, Vessella RL. 2003. "Detection and isolation of prostate cancer cells from peripheral blood and bone marrow". *Urology* 61:277-81.

90 Nam JM, Stoeva SI, Mirkin CA. 2004. "Bio-bar-code-based DNA detection with PCR-like sensitivity". *J. Am. Chem. Soc.* 126:5932-33.

91 Nam JM, Thaxton CS, Mirkin CA. 2003. "Nanoparticle-based bio-bar codes for the ultrasensitive detection of proteins". *Science* 301:1884-86.

92 Wang WU, Chen C, Lin KH, Fang Y, Lieber CM. 2005. "Label-free detection of small-molecule-protein interactions by using nanowire nanosensors". *Proc. Natl. Acad. Sci. USA* 102:3208-12.

93 Cui Y, Wei QQ, Park HK, Lieber CM. 2001. "Nanowire nanosensors for highly sensitive and selective detection of biological and chemical species". *Science* 293:1289-92.

94 Hahm J, Lieber CM. 2004. "Direct ultrasensitive electrical detection of DNA and DNA sequence variations using nanowire nanosensors". *Nano Lett.* 4:51-54.

95 Zheng GF, Patolsky F, Cui Y, Wang WU, Lieber CM. 2005. "Multiplexed electrical detection of cancer markers with nanowire sensor arrays". *Nat. Biotechnol.* 23:1294-301.

96. Wong SS, Joselevich E, Woolley AT, Cheung CL, Lieber CM. 1998. "Covalently functionalized nanotubes as nanometre-sized probes in chemistry and biology". *Nature* 394:52-55.

97 Woolley AT, Guillemette C, Cheung CL, Housman DE, Lieber CM. 2000. "Direct haplotyping of kilobase-size DNA using carbon nanotube probes". *Nat. Biotechnol.* 18:760-63.

98 Wang ZH, Liu J, Liang QL, Wang YM, Luo G. 2002. "Carbon nanotube-modified electrodes for the simultaneous determination of dopamine and ascorbic acid". *Analyst* 127:653-58.

99 Wang J, Liu GD, Jan MR. 2004. "Ultrasensitive electrical biosensing of proteins and DNA: carbon-nanotube derived amplification of the recognition and transduction events". *J. Am. Chem. Soc.* 126:3010-11.

100 Gref R, Minamitake Y, Peracchia MT, Trubetskoy V, Torchilin V, Langer R. 1994. "Biodegradable long-circulating polymeric nanospheres". *Science* 263:1600-3.

101 Ringsdorf H. 1975. "Structure and properties of pharmacologically active polymers". *J. Polym. Sci. Polym. Symp.* 51:135-53.

102 Davis FF. 2002. "The origin of pegnology". *Adv. Drug Deliv. Rev.* 54:457-58.

103 Moghimi SM, Hunter AC. 2000. "Poloxamers and poloxamines in nanoparticle engineering and experimental medicine". *Trends Biotechnol.* 18:412-20.

104 Park EK, Lee SB, Lee YM. 2005. "Preparation and characterization of methoxy poly(ethylene glycol)/poly(epsilon-caprolactone) amphiphilic block copoly-meric nanospheres for tumor-specific folate-mediated targeting of anticancer drugs". *Biomaterials* 26:1053-61.

105 Chari RV. 1998. "Targeted delivery of chemotherapeutics: tumor-activated pro-drug therapy". *Adv. Drug Deliv. Rev.* 31:89-104.

106 Mansour AM, Drevs J, Esser N, Hamada FM, Badary OA, *et al.*, 2003. "A new approach for the treatment of malignant melanoma: enhanced antitumor ef-ficacy of an albumin-binding doxorubicin prodrug that is cleaved by matrix metalloproteinase 2". *Cancer Res.* 63:4062-66.

107 Guo X, Szoka FC. 2003. "Chemical approaches to triggerable lipid vesicles for drug and gene delivery". *Acc. Chem. Res.* 36:335-41.

108 Allen TM. 2002. "Ligand-targeted therapeutics in anticancer therapy". *Nat. Rev. Cancer* 2:750-63.

109 Kannagi R, Izawa M, Koike T, Miyazaki K, Kimura N. 2004. "Carbohydrate-mediated cell adhesion in cancer metastasis and angiogenesis". *Cancer Sci.* 95:377-84.

110 Yamazaki N, Kojima S, Bovin NV, Andre S, Gabius S, Gabius HJ. 2000. "Endogenous lectins as targets for drug delivery". *Adv. Drug Deliv. Rev.* 43:225-44.

111 Bennis S, Chapey C, Couvreur P, Robert J. 1994. "Enhanced cytotoxicity of doxorubicin encapsulated in polyisohexylcyanoacrylate nanospheres against multidrug-resistant tumor cells in culture". *Eur. J. Cancer* 30A:89-93.

112 Larsen AK, Escargueil AE, Skladanowski A. 2000. "Resistance mechanisms associated with altered intracellular distribution of anticancer agents". *Pharmacol. Ther.* 85:217-29.

113 Links M, Brown R. 1999. "Clinical relevance of the molecular mechanisms of resistance to anticancer drugs". *Expert Rev. Mol. Med.* 1999:1-21.

114 Krishna R, Mayer LD. 2000. "Multidrug resistance (MDR) in cancer—mechanisms, reversal using modulators of MDR and the role of MDR modulators in influencing the pharmacokinetics of anticancer drugs". *Eur. J. Pharm. Sci.* 11:265-83.

115 Vauthier C, Dubernet C, Chauvierre C, Brigger I, Couvreur P. 2003. "Drug delivery to resistant tumors: the potential of poly(alkyl cyanoacrylate) nanopar-ticles". *J. Control. Release* 93:151-60.

116 De Verdiere AC, Dubernet C, Nemati F, Soma E, Appel M, *et al.*, 1997. "Reversion of multidrug resistance with polyalkylcyanoacrylate nanoparticles: towards a mechanism of action". *Br. J. Cancer* 76:198-205.

117 Blagosklonny MV. 2003. "Targeting cancer cells by exploiting their resistance". *Trends Mol. Med.* 9:307-12.

118 Tsuruo T. 2003. "Molecular cancer therapeutics: recent progress and targets in drug resistance". *Intern. Med.* 42:237-43.

119 Leamon CP, Reddy JA. 2004. "Folate-targeted chemotherapy". *Adv. Drug Deliv. Rev.* 56:1127-41.

120 Leamon CP, Low PS. 2001. "Folate-mediated targeting: from diagnostics to drug and gene delivery". *Drug Discov. Today* 6:44-51.

121 Stella B, Arpicco S, Peracchia MT, Desmaele D, Hoebeke J, *et al.*, 2000. "Design of folic acid-conjugated nanoparticles for drug targeting". *J. Pharmaceut. Sci.* 89:1452-64.

122 Merisko-Liversidge E, Liversidge GG, Cooper ER. 2003. "Nanosizing: a formulation approach for poorly-water-soluble compounds". *Eur. J. Pharm. Sci.* 18:113-20.

123 Fishman ML, Cooke PH, Coffin DR. 2004. "Nanostructure of native pectin sugar acid gels visualized by atomic force microscopy". *Biomacromolecules* 5:334-41.

124 Diaz JF, Strobe R, Engelborghs Y, Souto AA, Andreu JM. 2000. "Molecular recognition of Taxol by microtubules—kinetics and thermodynamics of binding of fluorescent Taxol derivatives to an exposed site". *J. Biol. Chem.* 275:26265-76.

125 Nicolaou KC, Riemer C, Kerr MA, Rideout D, Wrasidlo W. 1993. "Design, synthesis and biological-activity of protaxols". *Nature* 364:464-66.

126 Singla AK, Garg A, Aggarwal D. 2002. "Paclitaxel and its formulations. *Int. J. Pharm.* 235:179-92.

127 Albumin-bound paclitaxel (Abraxane) for advanced breast cancer". 2005. *Med. Lett. Drugs Ther.* 47:39-40.

128 Garber K. 2004. "Improved paclitaxel formulation hints at new chemotherapy approach". *J. Natl. Cancer Inst.* 96:90-91.

129 Gelderblom H, Verweij J, Nooter K, Sparreboom A. 2001. "Cremophor EL: the drawbacks and advantages of vehicle selection for drug formulation". *Eur. J. Cancer* 37:1590-98.

130 Vicent MJ, Duncan R. 2006. "Polymer conjugates: nanosized medicines for treating cancer". *Trends Biotechnol.* 24:39-47.

Related Resources

Ferrari M. 2005. "Cancer nanotechnology: opportunities and challenges". *Nat. Rev. Cancer* 5:161-71.

Yezhelyev M, Gao XH, Xing Y, Al-Hajj A, Nie SM, O'Regan RM. 2006. "Emerging use of nanoparticles in diagnosis and treatment of breast cancer". *Lancet Oncol.* 7:657-767.

Sinha R, Kim GK, Nie SM, Shin DM. 2006. "Nanotechnology in cancer therapeutics: bioconjugated nanoparticles for drug delivery". *Mol. Cancer Ther.* 5:1909-17.

Xing Y, Smith AM, Agrawal A, Ruan G, Nie SM. 2006. "Pro.ling single cancer cells and clinical tissue specimens with semiconductor quantum dots". *Int. J. Nanomed.* 1:473-81.

Gao XH, Yang L, Petros JA, Marshall FF, Simons JW, Nie SM. 2005. "In-vivo molecular and cellular imaging with quantum dots". *Curr. Opin. Biotechnol.* 16:63--72.

National Cancer Institute (NCI) Alliance for Nanotechnology in Cancer. *http://nano.cancer.gov*

Section III

Progress and Future Outlook

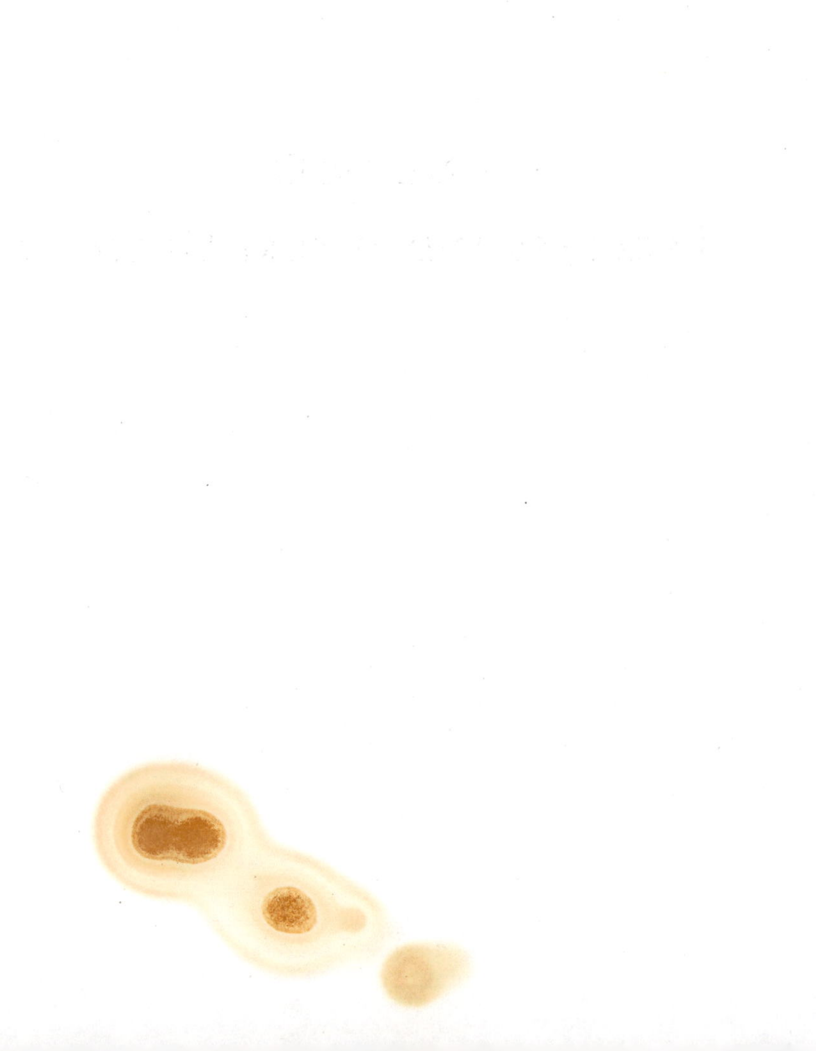

9

Change Required for the National Nanotechnology Initiative as Commercialization Eclipses Discovery

Matthew M Nordan

The National Nanotechnology Initiative (NNI) is a great success; it has not only funded prodigious fundamental research, but has also catalyzed a virtuous cycle of innovation manifested in corporate R&D and venture capital. However, The landscape is different today than when the NNI commenced in 2001. Nanotech's discovery phase has given way to commercialization – tens of billions of dollars worth of products now incorporate nanotech – and other nations are eroding the US's dominant position. As the NNI is reauthorized, its focus should shift to application development and manufacturing scale-up – and its approach to environmental, health, and safety (EHS) issues must be overhauled.

The NNI has Catalyzed a Virtuous Cycle of Innovation

Nanotechnology is the purposeful engineering of matter at scales of less than 100 nanometers (nm) to achieve size-dependent properties and functions. Nanotech is not a new industry or market, but rather an enabling set of technologies that impact a wide variety of industries through a nanotech value chain. This value chain starts with nanomaterials like carbon nanotubes and dendrimers, which are incorporated into intermediate products like memory chips and drug delivery systems, which are in turn used to make enhanced final goods like mobile phones and cancer therapies (see Figure 1). Lux Research projects that new, emerging nanotechnology applications will affect nearly every type of manufactured product through the middle of the next decade, becoming incorporated into 15% of global manufacturing output totaling $2.6 trillion in 2014 (see Figures 2 and 3).[1]

Figure 1: Nanotech adds Value across Industry Value Chains in Three Stages

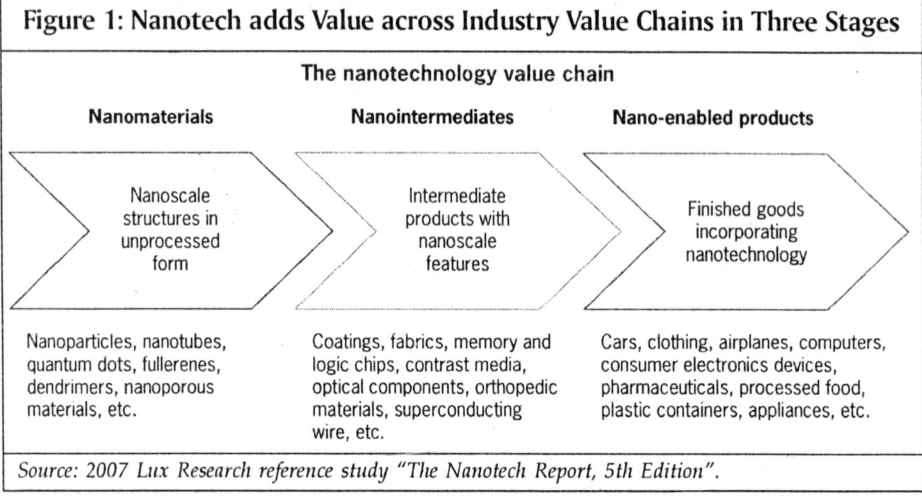

Source: 2007 Lux Research reference study "The Nanotech Report, 5th Edition".

Introduced in 2001 and signed into law in 2003, the US National Nanotechnology Initiative is the federal government's coordinating program for publicly-funded nanotechnology research, which has inspired similar efforts in countries worldwide from Germany to South Korea. By the core measure of scientific output, the NNI has been a great success – US scientists have published 55,661 journal articles on nanoscale science and engineering since 2001, 27% of the world's total. But in addition to this, the very presence of the NNI has catalyzed a virtuous cycle of innovation, manifested in:

[1] Source: October 2004 Lux Research report "Sizing Nanotechnology's Value Chain."

Figure 2: Product Categories are Incorporating Emerging Nanotechnology at Different Rates

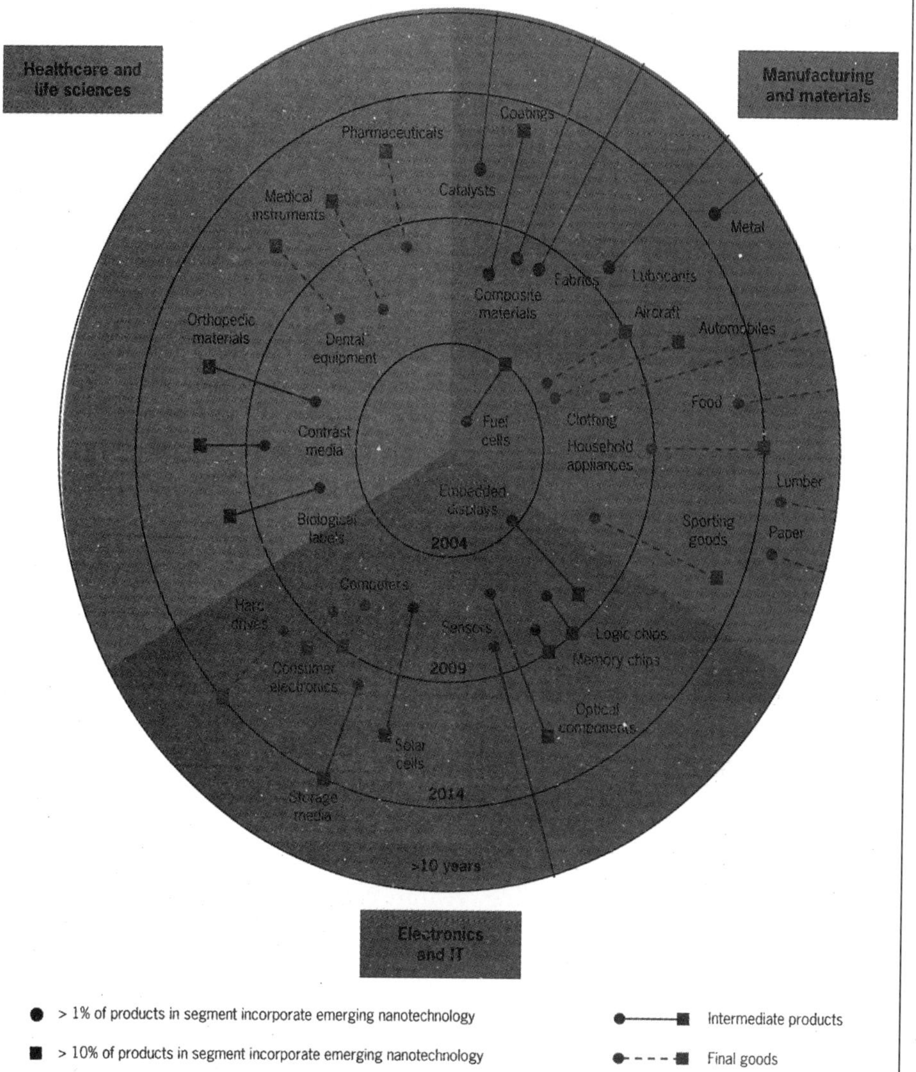

● > 1% of products in segment incorporate emerging nanotechnology

■ > 10% of products in segment incorporate emerging nanotechnology

●——■ Intermediate products

●- - - -■ Final goods

Source: 2007 Lux Research reference study "The Nanotech Report, 5th Edition".

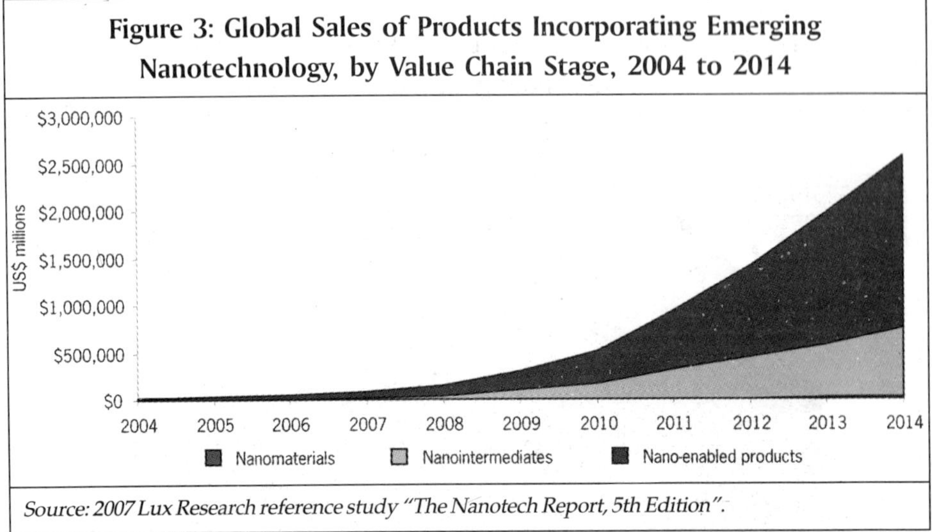

Figure 3: Global Sales of Products Incorporating Emerging Nanotechnology, by Value Chain Stage, 2004 to 2014

Source: 2007 Lux Research reference study "The Nanotech Report, 5th Edition".

- **Corporate R&D Spending.** Large US corporations from GE to Motorola spent \$2.4 billion in nanotechnology R&D in 2007, up 22% from 2006 and 557% from 2000, the year before the NNI's introduction. The 2007 figure was 23% higher than US government nanotechnology funding at the federal and state level combined.[2] These efforts include in-house research like GE's Nanotechnology Advanced Technology Program, broad collaborations like Cabot Corporation's Fine Particle Network, and joint ventures like DA Nanomaterials, created by DuPont and Air Products. Without the NNI as a widely-publicized focusing mechanism for nanotechnology research, it's unlikely that this intense corporate focus on nanoscale science and engineering would have materialized.
- **Venture Capital (VC) Funding.** Venture capitalists are always on the lookout for compelling investment themes, as well as non-dilutive sources of financing that can help sustain the companies they invest in through notoriously rocky early stages. The NNI has provided both, serving as a validator that has helped open VCs' wallets to materials science investments in a fashion never before seen. In 2007 VC firms put \$632 million into US-based nanotech start-ups in 2007, more than four times the figure in the year before the NNI was initiated (see Figure 4).[3]

2 Source: 2007 Lux Research reference study "The Nanotech Report, 5th Edition."

3 Source: March 2008 Lux Research report "How Venture Capitalists Are Misplaying Nanotech."

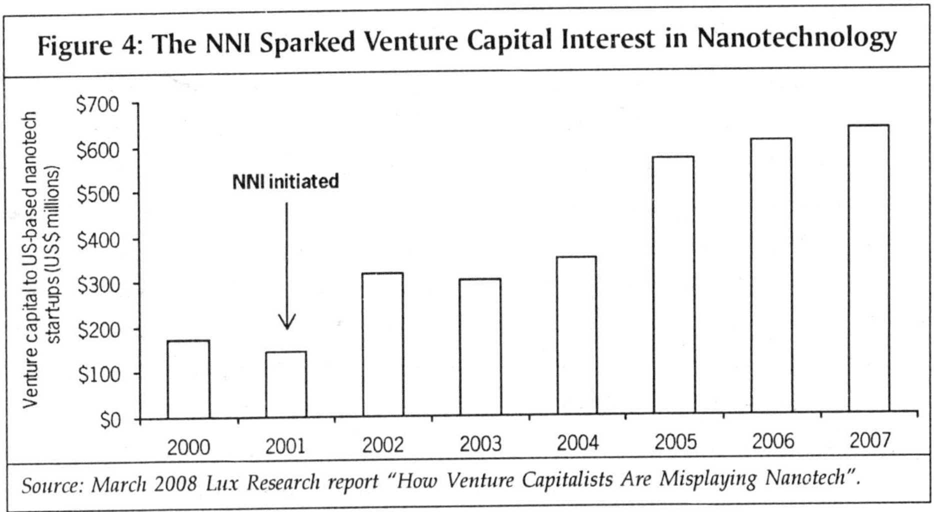

Figure 4: The NNI Sparked Venture Capital Interest in Nanotechnology

Source: March 2008 Lux Research report "How Venture Capitalists Are Misplaying Nanotech".

- **New Companies and New Jobs.** Consider A123Systems, which uses nanostructured lithium iron phosphate electrodes to make advanced batteries now being evaluated for use in electric vehicles like GM's Chevy Volt. In the mid-1990s, the Arsenal complex in the city of Watertown, Massachusetts was 750,000 square feet of empty, crumbling space. Now, A123Systems is its biggest tenant, commercializing battery devices based on research by Yet-Ming Chiang at MIT – precisely the type of research that the NNI funds. In just four years, A123 has gone from a few dozen employees to more than 1,000 – and helped to shift the center of battery innovation from east Asia to the United States.

The Nanotech Landscape is very Different Today than when the NNI Launched

When the NNI took shape in 2001, nanotechnology activity focused on early-stage laboratory research with little commercial impact, and the US was alone in the world in having a nationwide coordinating program for nanotech. Today, both factors have changed. Nanotechnology has shifted from its discovery phase into its commercialization phase – and at the same time, the dominant competitive position of the United States has been eroded by other nations.

Nanotech Commercialization is Eclipsing Discovery

In the last seven years, emerging nanotechnology has increasingly become a fact of life and of business, as the technology has shifted from an era of *discovery* to one of *commercialization*. In this fashion, nanotechnology follows the example of other world-changing technologies like polymer science and biotechnology. For these emerging technologies, everything starts with the discovery phase – a period of basic research and application development – which has a characteristic time span, give or take a bit, of about 20 years. It's then that a tipping point gets reached, triggering the commercialization phase – where the technology's long-term impact is manifested (see Figure 5).

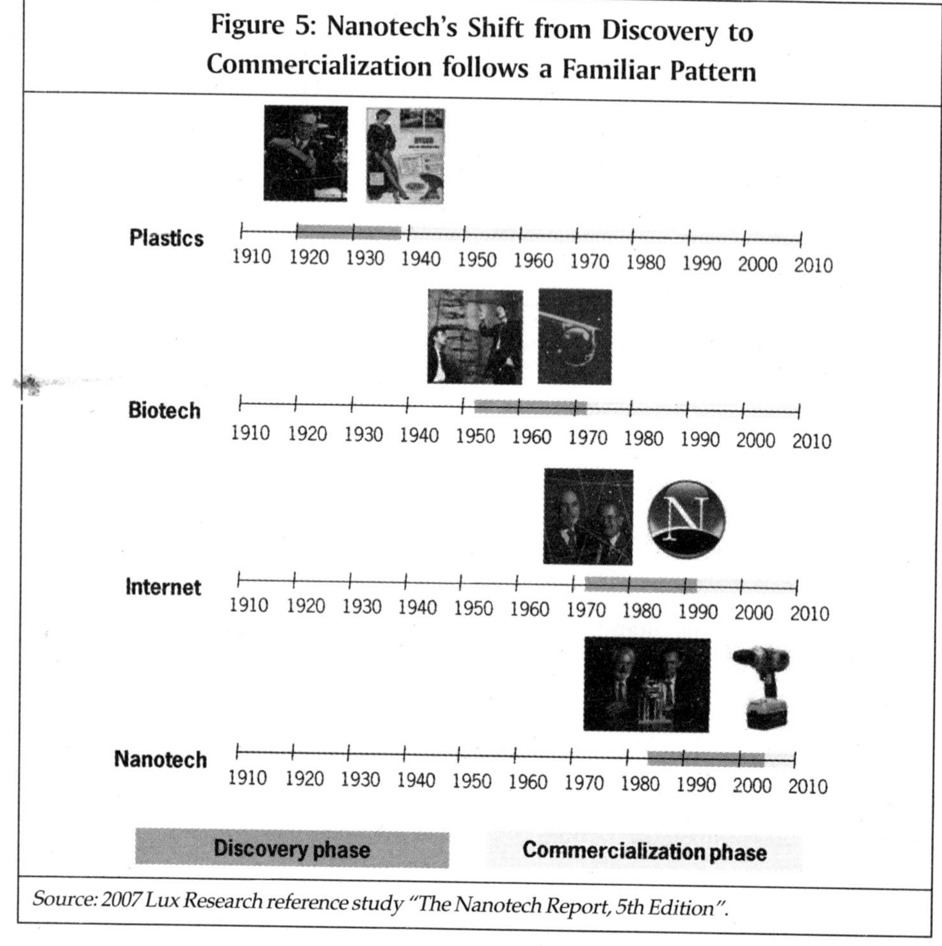

Figure 5: Nanotech's Shift from Discovery to Commercialization follows a Familiar Pattern

Source: 2007 Lux Research reference study "The Nanotech Report, 5th Edition".

For instance, plastics' discovery phase started in the 1920s, when scientist Wallace Corrothers at DuPont started working work on synthesizing nylon. In 1937, he was issued his patent on the material. Two years afterward – about 20 years after discovery began – American women bought 64 million pairs of nylon stockings; once the commercialization threshold was reached, it took off fast. In biotechnology, James Watson and Francis Crick characterized DNA in 1953, and 20 years later, right on cue, Stanley Cohen and Robert Boyer applied genetic engineering techniques to synthesize insulin for the first time. Genentech, the first biotech start-up, was founded in 1976, and commercialization has since skyrocketed: In 2006, revenues of publicly-traded biotech companies topped $65 billion. In information technology, Vint Cerf and Robert Kahn proposed the Internet protocol in 1974. The number of Internet users grew gradually to the single-digit millions up through 1993, but began to skyrocket in about 1994, the year Netscape's browser was released – reinventing communication and commerce in the process.

Nanotech's discovery phase started in the mid-1980s with the invention of scanning probe microscopes that enabled scientists to visualize matter at the nanoscale for the first time. Innovations have reached the market in electronics, as A123Systems' nanostructured battery electrodes appeared on store shelves in Black & Decker's Dewalt line of power tools; in healthcare, as nanoparticulate drug reformulations like Abbott's cholesterol drug Tricor have found their way into doctors' repertoire; and in materials and manufacturing, as PPG's coatings have improved the performance of millions of automobiles. According to our research, approximately $88 billion in manufacturing output worldwide incorporated emerging nanotechnology in 2007.

The Dominant Position of the US is being Eroded

Each year, Lux Research conducts an annual assessment of international competitiveness in nanotechnology, ranking 19 nations worldwide on their nanotechnology activity and technology commercialization strength. On an absolute basis, the US remains the world leader in nanotech. Two factors, however, should give US policymakers pause:

- **The US does *not* Lead on a Relative Basis.** Relative to our population and the size of our economy, the US pales in comparison to other countries when it comes to nanotechnology activity. For example, when government funding is considered on an absolute basis, the US topped the charts in 2007. However, when the same figures are considered on a per capita basis at purchasing power parity, the US takes eighth place, with funding half that of Taiwan, and behind Germany, Sweden, and France (see Figure 6).[4]

Figure 6: US Government Nanotech Funding Lags other Nations on a Relative Basis

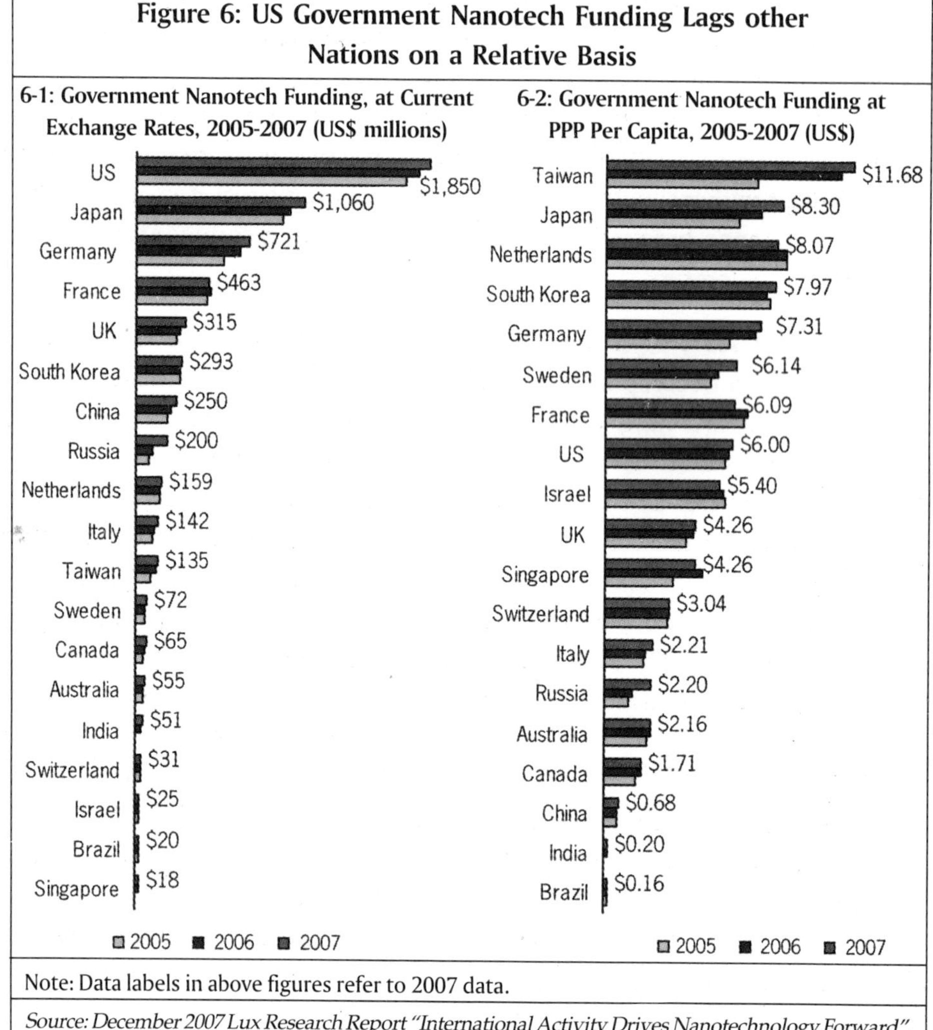

Note: Data labels in above figures refer to 2007 data.

Source: December 2007 Lux Research Report "International Activity Drives Nanotechnology Forward".

- **Other Countries are Catching Up.** Since we began performing our international competitiveness rankings in 2005, the position of the US has remained static while other countries have vaulted upwards in their nanotechnology activity (see Figure 7). For example, nanotech funding is growing in the EU at twice the rate of the United States, by putting the EU on track to claim the mantle of nanotechnology leadership due to a renewed focus on nanoscale science and engineering in the 7th Framework Programme for research. Russia recently funded a state nanotechnology corporation with $5 billion of public financing. And scientists in China published nearly as many scientific journal articles on nanoscale science

Figure 7: Other Nations are Gaining on the US in Nanotechnology

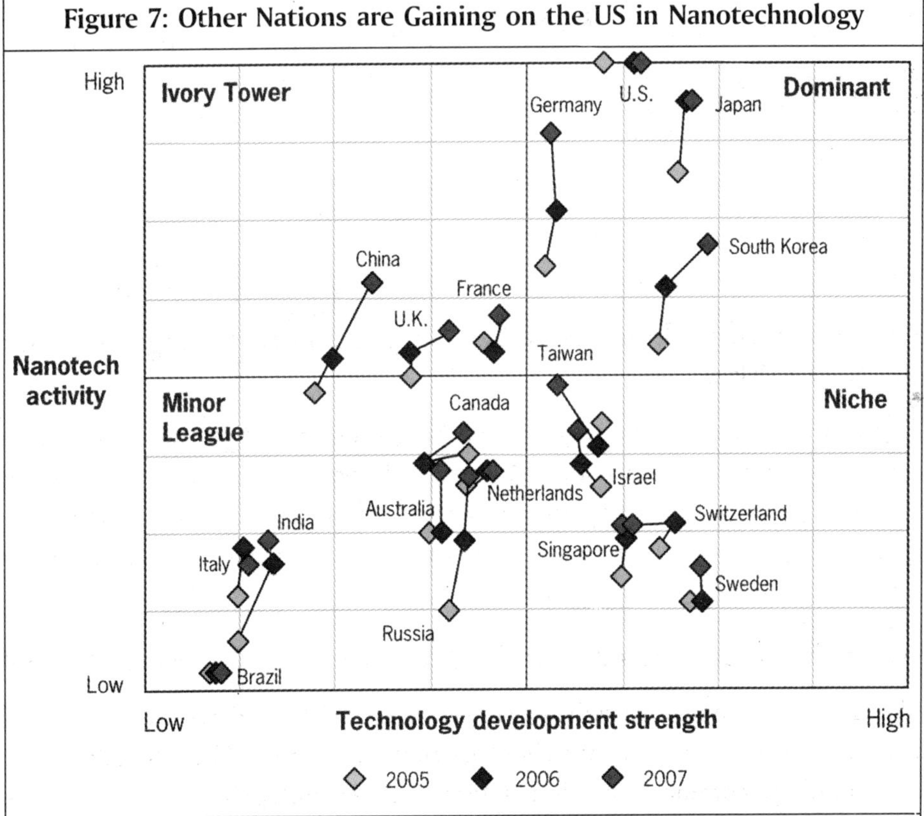

Source: December 2007 Lux Research report "International Activity Drives Nanotechnology Forward".

and engineering in 2007 as those the US did, at 7,282 to 7,528 (see Figure 8). While the quality of these publications has been suspect in the past, the citation rate of nanotech journal articles from China – a measure of their quality – has doubled in the last decade.[5]

Figure 8: Journal Articles on Nanoscale Sciences and Engineering Topics

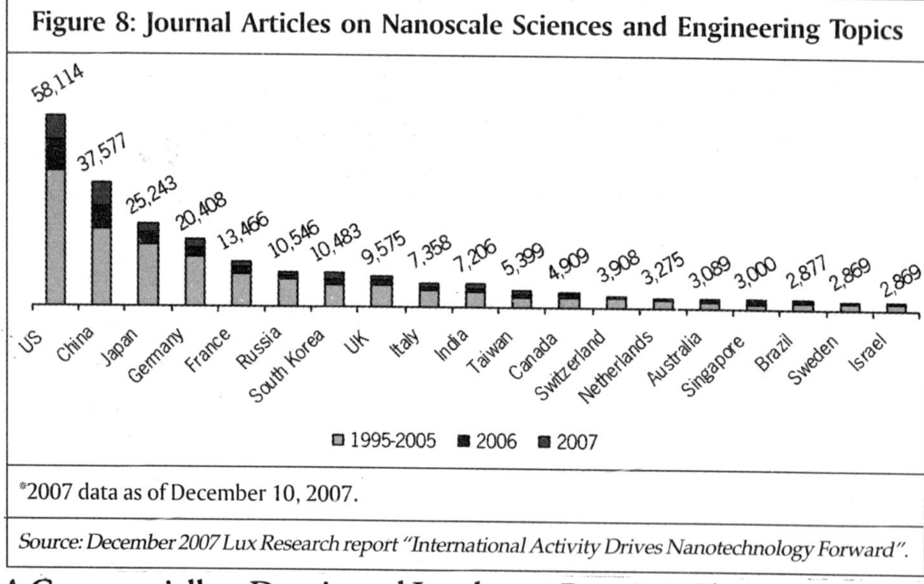

*2007 data as of December 10, 2007.

Source: December 2007 Lux Research report "International Activity Drives Nanotechnology Forward".

A Commercially – Dominated Landscape Requires Change to unlock the NNI's Value

Clearly, the NNI should be reauthorized. But in the context of growing nanotech commercialization and increased international competitiveness, the onus is on Congress to eliminate roadblocks to market introduction for nanotechnology applications. Most of these changes are not specific to nanotech, although a few key ones are.

Nanotech's Pervasiveness Means that Most Required Changes are General

Many of the changes that will help transition NNI-funded research to market have nothing to do with nanotechnology specifically, but address broader issues in technology commercialization. Given nanotechnology's diversity, and the

5 Source: Science Citation Index as of December 10, 2007; search terms (country), (year), and (quantum dot OR nanostruc* OR nanopartic* OR nanotub* OR fulleren* OR nanomaterial* OR nanofib* OR nanotech* OR nanocryst* OR nanocomposit* OR nanohorn* OR nanowir* OR nanobel* OR nanopor* OR dendrimer* OR nanolith* OR nanoimp* OR nano-imp* OR dip-pen).

breadth of product categories that it touches, this is to be expected: As goes technology in general, so goes nanotech. These changes include:

- **Attracting US Students to Science and Engineering, and Retaining Foreign Ones.** Funding for nanotechnology R&D will amount nothing without a steady stream of trained scientists and engineers entering the workforce. The US should strengthen programs designed to inspire students with wonder for the physical sciences in K-12 and undergraduate education to nurture homegrown talent. But it should also reconsider the effect of visa tightening on the inflow of foreign science and technology graduate students, and expands H1-B visa programs to allow students that have earned advanced degrees in science and engineering in the US to remain here – rather than repatriating taking with them the skills they acquired in the US. The lesson of A123Systems is instructive: Had Yet Ming-Chiang returned to his native country, its 1,000 employees would likely be in Taiwan.
- **Reducing the Cost of doing Business for Start-ups Seeking Public Markets.** Start-up companies looking to make initial public offerings (IPOs) on the public markets face immense administrative costs to comply with regulations such as Sarbanes-Oxley. Easing these burdens will unshackle them. It's important to note that of the 14 nanotech start-ups that have gone public, most have done so on foreign exchanges where the cost of doing business is lower.
- **Introducing Financial Mechanisms to Encourage Collaboration between Small and Large Firms.** Nanotechnology commercialization has followed a pattern similar to biotech, in which small, innovative companies develop breakthrough technologies that incumbent corporations bring to market: Silver nanoparticle antimicrobial company Nucryst Pharmaceuticals relied on wound care dressing maker Smith & Nephew to get to market, while A123Systems found its partner in Black & Decker. Congress can grease the wheels of nanotechnology commercialization by creating financial mechanisms that help small firms to collaborate effectively with large ones. One example of such a measure would be enabling small companies to transfer their net operating losses to their corporate partners – allowing those partners to reap the tax benefits of research investments, which the loss-making smaller companies can't claim.

A Few Nanotech-specific Changes are Necessary

In addition to these general reforms, a smaller number of changes specific to nanotech are also required. There are two specific actions we think Congress should take now:

- **Shift some of NNI's Focus to Application Development and Manufacturing Scale-up.** A reauthorized NNI should focus on not just on basic research, but also on precompetitive R&D into nanomaterials application development and manufacturing scale-up. Currently the technical challenges that are limiting nanotech commercialization are not as much in synthesizing nanomaterials or understanding their fundamental properties as in learning how to integrate them into products and manufacturing them economically in large volumes. The Department of Energy's Nanomanufacturing initiative, run out of its Industrial Technology Program, is a model case study – it aims to introduce shared Nanomanufacturing centers as pilot facilities on the model of the NNI's existing user centers for nanoscale analytical equipment.
- **Take a Completely Different Approach to Environmental, Health, and Safety (EHS) Issues.** In our work with companies looking to take advantage of nanotech innovation, the single concern that comes up more than any other is potential EHS risks of nanomaterials. While it's of course incumbent on companies developing nano-enabled products to test their own products to ensure safety, there's an important role for government to play in resolving these concerns – by funding basic research on nanomaterials EHS risk that no individual firm can afford, and by establishing clear regulatory guidelines for companies working with nanomaterials.

  On the first point, the NNI should be generously funding basic research on the EHS risks of nanomaterials – just as NNI-funded research on nanoscience has supported the deployment of real-world nanotech applications, the results of NNI-funded EHS work would help companies complete their own EHS evaluations. Unfortunately, funding levels remain too low to have the desired impact, and, even more critically, the NNI has never effectively addressed EHS issues surrounding nanotech with a comprehensive, interagency plan for required EHS research. The Nanotechnology Environmental and Health

Implications (NEHI) Working Group report on EHS issues has not filled this gap – it fails to prioritize specific materials and applications for research, avoiding the trade-offs that are inherent in any meaningful strategy – and the EPA's internal review of its own nanomaterials EHS activities, by definition, does not cross agencies. This lack of coordination is hampering development, and must change. The best way to move forward on this front would be to execute the nanomaterials EHS strategy by the National Academies' Board on Environmental Studies and Toxicology – Congress has already appropriated funds for this study, but the work has not yet been started.

Second, ambiguity surrounding how nanomaterials will be regulated must be dispelled. It is still not clear how current regulations apply to nanoparticles or whether and when agencies will issue new ones – leaving firms that work with nanomaterials confused about how to plan for regulatory rulings. The companies we speak with are actually eager for appropriate regulatory guidance about nanomaterials, to ensure a level playing field and to help them guarantee the safety of workers, consumers, and the environment. While companies are generally pleased about how the EPA, for example, has communicated with them so far, they're also frustrated by how slow those agencies have been to set specific guidance, as witnessed by the glacial pace of the EPA's voluntary Nanoscale Materials Stewardship Program. Seven years after the NNI's launch, it's still unclear to most commercial entities when and how the materials they work with will be treated under the EPA's Toxic Substances Control Act – forming a real commercialization gating factor.

At Lux Research, we applaud the efforts that have taken place so far under the National Nanotechnology Initiative, which have made the US a world leader in nanotechnology and are bringing real economic benefits to our nation. We're confident that a renewed NNI, with adjustments like those outlined above, will increase these benefits – and enable nanotechnology to help address the challenges the country faces in combating disease, moving toward energy independence, and sustaining economic growth.

(Matthew M Nordan, The author can be reached at matthew.nordan@luxresearchinc.com).

10

Nanotechnology for Development or Knowledge Enclaves?
The World Bank Case for Latin America

Guillermo Foladori, Mark Rushton and Edgar Zayago Lau

A great deal has been written in recent years about nanotechnology, its revolutionary significance for science and real-world applications that are touted as being capable of profoundly transforming the world in which we live. Yet, very little has been written about how they are incorporated into the context of the knowledge economy. In this article, the authors analyze the World Bank's intention to develop Scientific Millenium Initiatives as Centers of Excellence in Latin America to boost competitiveness and encourage economic growth, which is understood by the World Bank as a requirement for development. Nanotechnology is a strategic area within these projects. However, the authors conclude that rather than leading to development, these centers are more likely to become knowledge enclaves with little impact on the real development challenges of the region.

Source: http://estudiosdeldesarrollo.net/, Latin American Nanotechnology & Society Network (RELaNS)(2007).

Introduction

A great deal has been written in recent years about nanotechnology, its revolutionary significance for science and real-world applications that are touted as being capable of profoundly transforming the world in which we live. Little, however, has been published that situates nanotechnology within the broad context of development. One approach to place it in the context is an examination of nanotechnology as part of the "Knowledge Economy" paradigm. Our interest lies in exploring that subject and how it is applied to the context of Latin America. In part one, the concepts with which we examine how this new industry is portrayed by its supporters and those who call for a deeper analysis of its adoption are presented. In part two, some aspects of the historical context and some relevant theoretical underpinnings of "development" are offered, as a foundation for analyzing nanotechnology's status as a key development mechanism according to the World Bank's prescriptions. In part three, the global trade regime as it pertains to intellectual property and patent protection in the context of the use of knowledge to aid economic growth is discussed. Part four looks at Latin America's experience with nanotechnology development under the World Bank's guidance, and examines the resources being committed to the development of "knowledge enclaves." Also examined some of the challenges faced by the countries as they faced attempt to develop a top-down high technology approach in areas where infrastructure and educational base are less than ideal.

1. Nanotechnology and the Question of Development

The "development debate" has existed for decades, with definitions offered, discarded and reformulated. Whereas the generation of profit and economic success lay at the core of much institutional development policy, today most development theorists and practitioners favour a wider-ranging perspective that includes the environment, gender, labour, culture and various other related aspects of societal change, focusing upon the improvement of the material quality of life for all citizens. This is the window through which the question of nanotechnology should be examined.

What, then, is an appropriate definition of development as it pertains to the question of nanotechnology? The presently emerging revolution in the nanosciences and corresponding nanotechnology represents, perhaps, one of the most profound technological revolutions humankind has experienced, with great potential for altering at the atomic level the very properties of matter.

In describing this new industry, proponents often cite the potential in particular for the developing countries to embrace nanotechnology as an excellent solution for countless problems, ranging from safe water; energy production; and health care. A core tenet of the nanotech revolution is the potential for significant new economic opportunities, and since developing nations are poor, nanotechnology is thus seen as a way to "catch up" if only they are able, or assisted, to take advantage of the technology quickly, to jump aboard the ship before it sails out of reach. In reviewing the debates on nanotechnology development, we noticed that the University of Toronto Joint Centre for Bioethics (UTJCB) and the Task Force on Science, Technology and Innovation of the United Nations Millennium Project talk about the potential benefits of this revolution for development, but are scant on the details, proposing a technical solution without – it seems – being sufficiently conscious of the broader problems of development (Salamanca-Buentello, *et al.,* 2005; Juma & Yee-Cheong, 2005).[1] Those authors could be identified as having a *technology-neutral* approach, as they view underdevelopment as the result of a lack of technology, and thus nanotechnology is portrayed as a solution. Another perspective is that of the ETC Group (2003) and Invernizzi & Foladori (2005), these authors represent a *contextual* approach as they favour an analysis of the socioeconomic context parallel to influencing technology development. In addition, they understand development as a socioeconomic problem and technology as no more than a tool subsumed to economic trends.

Nanotechnology is certainly a path which represents great potential. We argue, however, that this potential is far from certain to be realized without due attention paid to the big picture. Technological advances, however revolutionary, do not guarantee better living conditions for the poor and workers. Even in those contexts where nanotechnology could be fully integrated in Least Developed Countries (LDCs) national development plans, socio-economic structures are unlikely to change as a consequence in the absence of progressive planning to control and

1 A chronological survey of positions on nanotechnology and developing countries can be found in Invernizzi, Foladori and Maclurcan (2007).

push mechanisms in favor of peoples needs. Rather, nanotechnology may further the technological and socio-economic isolation of the poor, worsening existing gaps, despite the profits generated.

2. The Knowledge Economy and its Significance for Development

The global race for industrial development is increasingly dependent and based upon a process of "productive transformation," – technological conversion of the industrial production apparatus – In this context, nanotechnology, as the next industrial revolution, presumably provides fertile ground to cultivate the development process in LDCs. However, in order to integrate nanotechnology as a tool for development we need to understand the different theoretical frameworks surrounding the notion of development. The discussion about what development is or is not, is wide and associated with several theoretical currents. As a way of synthesis we can define some of the most important positions regarding the different notions of development.

One of the most prevalent and the first conceptual framework of development comes from the work of WW Rostow. His concept of development was directly linked to economic growth, modernization and industrialization, with high mass consumption as the indicator of full progress; presumably LDCs will have to pass through five stages of development to industrialize their economies: *Traditional, Transitional, Take-Off, Drive to Maturity* and *High Mass Consumption* (Rostow, 1960).

Raul Prebisch elaborated the center-periphery concept, using the analysis of the deteriorating terms of trade, the result of the quotient between the export price index and the import price index. By the mid 20th century the terms of trade deteriorated for countries that exported raw and basic materials in Latin America due to monopolization and unequal relations of power (with the exception of oil). This had a decidedly negative impact upon the industrialization process in those countries. From this analysis, Prebisch concluded that the problems of underdevelopment in Latin America have structural origins (Prebisch, 1950, 1984).[2]

[2] From the 1970s-on, terms of trade for commodities also experienced a decline.

In contrast to these positions, Paul A. Baran (1957), argued that the development of underdevelopment in LDCs was perpetuated by the lack of distribution of power among classes, the control over the economic surplus in all its forms and the inability of LDCs to compete with the advanced capitalist countries.

The current neoliberal process of economic growth that prevails in the world was based upon the working ideas of neoclassical economic theory. First Hayek (1944) and latter Friedman (1962, 1980), argued that the liberalization of trade and the integration of national economies are preconditions to encourage economic growth; as long as they are willing and able to successfully compete in that market, under rules of engagement that they have no ability to influence. In fact, during the military dictatorship in Chile, this theory was put into practice to presumably encourage its economic growth (Cypher, 2005; Valdez, 1989).

The idea that development equals economic growth has been contested by proponents of a broader definition of development. Within this framework it is believed that increased incomes are a means to achieve development but they will never be the end unto themselves (Sen, 1988; Streeten, 1981). The United Nations Development Programme (UNDP) in its 1990 report created a more comprehensive definition of what human development is: *a process of enlarging people's choices* (UNDP, 1990). Through time the notion of development has gone beyond economic parameters to incorporate issues of environment, gender, ethnicity and livelihoods (Ahooja-Patel, 1982; Chambers and Conway, 1998, 1995 and Chambers, 1987).

Nevertheless, it is important to point out that the hegemonic idea of what is development is still defined within an economic framework. On the agenda of development agencies at the international level, such as the World Bank and the IMF, economic performance remains the core objective of policy prescriptions. There is a recognition of the "incidentals" of development, including impacts of progress on culture and society, the environment, labour and the role of government, but these do not distract from the economic focus that is "development" for these agencies:

- The Bank has sharpened its support for the development agenda through a two-pillar strategy for reducing poverty that is based on building the climate for investment, jobs, and sustainable growth and on investing on poor people and empowering them to participate in development (World Bank, 2005).

In addition, within the framework of development as economic growth, the increase in the production of high technology products within the developed economies has been evident since the beginning of the 1990s and led to the widespread application of the term "Knowledge Economy" (or "Information Society") to refer to an economy in which innovation and knowledge is the driving force.

The World Bank ranks countries according to the share of high technology products in total exports. "High technology products" are considered to be those which are the results of intensive research and development (R&D), including aerospace, computers, pharmaceuticals, scientific instruments and electrical machinery. In 2004, for example, 34% of Ireland's exports were high-tech products; in South Korea 33%; in US 32%; but in Latin America, Chile exported only 5%; Brazil 12%; and México 21%[3] (World Bank, 2006). The World Bank analysis suggests that R&D has come to play an essential role in development, and this is the path that developing countries should follow to rise up from underdevelopment.

Organizations such as the World Bank, the OECD and the World Trade Organization (WTO) came to see knowledge and innovation as prerequisites for Third World countries in the process of development. The transformation of the industrial apparatus in LDCs now relies on the Knowledge Economy (World Bank, 1999).

With economic success weighing down the standard definition of development in the halls of the most influential development organizations in the world, efforts to orient nanotechnology toward serving the broader developmental interests of LDCs would appear to be an uphill battle. Can one expect that this technological revolution will have different results than those that came before? Or will nanotechnology simply be the latest method of creating profit that bypasses the

[3] In the case of México, the weight of maquiladora production and the strong intra-firm trade of US transnationals suggests that there is a need to be cautious in the analysis (Delgado Wise & Invernizzi, 2002).

interests of the majority of people who live in the developing world?

3. The Knowledge Economy and Developing Countries

Developed countries are, increasingly, exporting high technological products. At the end of the 20th century, the percentage of those exports was between 18 and 25% in the European Union, Japan and the United States – not including Ireland, with approximately 38% (UNDP 2006). The manufacture of high tech production is seen as the evidence of a transition to the knowledge economy.

In promoting the theoretical importance of knowledge, groups and organizations like the World Economic Forum and the World Bank created indices of Competitiveness for Growth and tables that rate different countries according to their position in the knowledge economy.[4]

In a capitalist world, the ability to bring development through the use of knowledge requires that it is oriented toward and in the service of businesses that can realize innovations to be transformed into commercial advantages. The Knowledge Economy implies an *education+innovation+competitivness* triad. This signifies a growing tendency toward the privatization of education and its placement in service to business. At the same time, business ventures needs to dedicate themselves increasingly to attain international competitiveness. The result is that the market, as the blind force behind competition, ends up orienting production. Although all current debates about innovation and the knowledge economy refer to "innovation with equity" or "competitiveness with sustainability" and similar terms, the fact is that all forms of economic planning remain regulated by unrestrained market forces. It remains to be seen over the decades to come whether economists who set development policies in LDCs will in fact be able to achieve the reduction of inequity and poverty.

How, then, can this dynamic be incorporated into countries that historically developed through primary material exports with scarce value-added? This is one of the issues that enthusiastic proponents of the knowledge economy must face. While in the developed countries the infrastructural conditions, training and

---

[4] See the World Economic Forum's Global Competitiveness Network: Global Competitiveness Report 2005-2006 at *http://www.weforum.org/en/initiatives/gcp/index.htm* [Last accessed 23 May 2007].

human skills in Science & Technology (S&T) have been built over decades, and in particular with attention given to the relatively recent technological revolutions in informatics and computers, biotechnology and telecommunications, in developing nations there do not exist the material infrastructure or the subjective conditions of professional training to embark upon the knowledge economy path. The "solution" that they have found is a top-down mechanism, or "knowledge enclave" plan. Developing countries could create "Centres of Excellence," institutes or research bodies with few researchers but significant resources, and with a strong relationship to industry.

During the 1980s, the World Bank concentrated its efforts in financial liberalization. A part of that orientation involved closing the Department of Science and eliminating the position of Scientific Advisor to the World Bank, embracing the idea of "free trade for development."

In 1994, the WTO instituted the worldwide patent regime TRIPS (Agreement on Trade-Related Aspects of Intellectual Property Rights) in order to guarantee patent protection of foreign trade operations. TRIPS sets down minimum standards for many forms of intellectual property regulation. TRIPS also established a legal system and mechanism for dispute settlement, including sanctions, for countries that do not comply with the legislation. Patent protection for new technology products permits the owner to set monopolistic prices for that product over a period of twenty years (WTO, 1994). In this way, the WTO with TRIPS ensures that large corporations that possess the majority of the patents have assured future monopoly profits.

The World Bank's *World Development Report 1998/99* carried the subtitle, *Knowledge for Development*, referring to the gap in knowledge between rich and poor countries (Masood, 1999). The basis for this change was the recognition that the economic liberalization of the 1980s had not attained the anticipated results, but rather had increased the gap between rich and poor countries, and increased their foreign debt. The Bank also indicated that the patent regimen had not resulted in the promotion of private research in areas of greatest impact on development. The mechanisms of the market were not sufficient to create incentives for research in areas with little return, for example, the "neglected diseases" or illnesses of the poor, such as treatments against malaria. In these cases, the Bank recommended that public funding should be directed towards subsidizing research (Nature, 1988); and it showed concern for the extension of intellectual property rights beyond products, to cover biotechnology achievements (Butler, 1998). According to the World Bank, the proposed course of action at the end of the 1990s was, then, to incorporate themes of innovation, S&T and the transfer of technology as key objectives of the Bank for developing countries.

4. Nanotechnology and "Knowledge Enclaves" in Latin America

Since the end of the 1990s, the World Bank and various other institutions have planned for the creation of a global network of "Millennium Scientific Initiatives." These will be the centres of excellence in developing countries, with the objective of promoting research in S&T under equal conditions of infrastructure and resources as exists in research centers in the developed countries (Macilwain, 1998).

The Chilean project was the prototype. In 1999, the government of Chile created the National Commission of Scientific Initiatives for the Millennium, with the objective of enhancing capacities in scientific research (DORCH, 1999); shortly thereafter, the World Bank provided a 5-million dollar loan for the first two-and-a-half year period, supplementing a 10-million dollar national budget (ICM, n/d a). The Millennium Scientific Institutes' (ICM) objectives were:

- ...to foster growth in scientific research capacities, employing and stimulating the best talent in the country, as a key factor for sustainable socio-economic development. The Programme anticipates that the creation of Centres of Scientific Excellence will give rise to Scientific Institutes and Scientific Nuclei under a competitive and transparent process. These centres will pursue scientific research on the frontier, the training of scientists and the establishment of links with the productive sector and other institutional agreements (ICM, n/d a).

Instead of adjusting the research lines to a particular national development plan or project, the Programme worked to identify talented Chileans within the country and beyond to drive research in the direction that they had an interest in pursuing. This science policy could appear elitist, but it was based on the idea that whatever the orientation of innovation, it would result in greater international competitivity and would guarantee development (understood as to winning space in the international market to promote the process of economic growth). Although terms such as "sustainable development," "combating inequity and poverty," and others that were aimed to humanize the concepts of innovation and competitivity were included in the formula, it is clear that the concept of development behind these sort of projects is based on the idea that the improvement of competitivity increases wealth in a country, and then such wealth would be automatically redistributed. Another way to view this issue is the argument that without an increase in capital there would be no possibility of distribution; and in any case, the policies of distribution of wealth, rather than those promoting innovation, are responsible for combating inequity and poverty. In this way, responsibilities are separated, but the reduction of poverty and inequity is directly linked to the increase in monetary supply.

Other objectives of the Chilean Millennium Initiative included attracting foreign talent and avoiding "brain drain." In this sense, the World Bank plan that Chile implemented as a pilot project, is *top-down*. Centres of Excellence were created for the most distinguished scientists, with the hope that they would facilitate alliances with private enterprise and lead to productive innovation. Although the spirit of the plan was to create the conditions for the researchers to stay in Chile and not migrate, it is debatable whether this could be achieved in enclaves of excellence with short support and without a concurrent basic educational reform effort to nourish and allow the replication, in the long-term, of a path for technological innovation. These enclaves of excellence would have to survive in a country where only 0.6 % of the GDP is destined for S&T; a very low figure, only a few decimals above México's commitment and clearly inferior to Brasil's (0.95%) in the Latin American context; it is certainly below the support provided in the developed countries and the ones that have adopted innovation in the last decades as a strategy for development, like South Korea (2.63 %); USA (2.68%); and México (0.41%) (OECD 2005) and the regional leader, Venezuela (2.11%).[5] Many other countries in the world followed variants of the

[5] "Inversión venezolana en ciencia alcanzó 2,11% del PIB" *http://www.scidev.net/News/index.cfm?fuseaction=readNews&itemid=3636&language=2* Last accessed 29 May 2007.

Chilean example. In Latin America, Mexico, Venezuela and Brazil established agreements with the World Bank to develop their own Millennium Initiatives.

Nanotechnology is considered one of the most important areas within contemporary innovation, and a paradigmatic example of research which must be supported in the transition into the knowledge economy. However, what matters in not the technological innovation per se; rather, it is the incorporation of that innovation in the manufacture of products with an international competitive advantage that makes the difference. In fact, competitiveness is one of the justifications – and in many cases the only one – for making use of public funds to research new technologies. The National Nanotechnology Initiative of the United States illustrates this idea, but it is also an important feature for other nanotechnology programs in countries such as Argentina and Brazil, and is present in reports issued by the Mexican and the Costa Rican Governments (Foladori, 2006); it is easily seen in the Malaysian Nanotechnology Centre or behind the official discourse of the Government of Thailand (Tun Razak, 2005; Tanthapanichakoon, 2005). Even though the spirit behind the discourse of competitiveness is to encourage development, and presumably, benefit society overall, the historic experience, which is the body sustaining that spirit, indicates the precise opposite. A country can improve its competitiveness without necessarily improving the living standards of its population, with the cost of increasing inequality, the Mexican case is a perfect illustration of this.

Nanotechnology innovation is an important aspect within the United Nations for developing countries. The *Task Force on Science, Technology and Innovation of the UN Millennium Development Project*, for example, released a report with the suggestive title *Innovation: Applying Knowledge in Development* (Juma & Yee-Cheong, 2005), where it put forward the idea that nanotechnology will be important to the developing world, because it requires little work, land and maintenance, is highly productive and cheap and requires only modest quantities of material and energy.

In the same vein, in February 2005, the International Centre for Science and the United Nations Industrial Development Organization organized a conference (*North-South Dialogue on Nanotechnology: Challenges and Opportunities*) specifically focused on the participation of developing countries in nanotechnology (Brahic, 2005a, 2005b; Brahic & Dickson, 2005). Representatives from governments, academia, international experts and representatives from industry took part. Of particular interest was the statement of the president of the Third World Academy of Sciences, Mohamed Hassan. He proposed the establishment of Centers of Excellence in Africa, thereby promoting cutting-edge S&T as necessary for developing countries to succeed (Hassan, 2005). The same idea has been discussed by the leaders of the world's most industrialized nations (Group of 8) since 2000, which explicitly backed the creation of Centers of Excellence in Africa to encourage the transfer and sharing of Science & Technology between developed and developing countries, during its annual summit in Scotland in 2005 (Dickson, 2005).

In Latin America, Brazil, Argentina and México are the countries where nanotechnology research have made particular advances (Foladori, 2006a). However, there are differences between their approaches. In 2001, Brazil introduced a national plan to scientific research networks with a one-million-dollar budget. Later, in 2004, it announced the Nanoscience and Nanotechnology Program, within the framework of the *Plan Pluri Anual de Desarrollo 2004-2007* (The Pluri Annual Plan for Development 2004-2007), for which the Brazilian government allocated 39-million dollars (MCT, 2004a, 2004b). Additionally, there are several funds from federal, provincial and international sources to sponsor nanotechnology research in Brazil. Most of these resources are centrally managed by the Ministry of Science and Technology in Brazil with the objective of advancing nanotechnology research.

The government of Argentina, on the other hand, created in 2005 the Argentinean Foundation for Nanotechnology with an estimated budget of 10-million dollars to cover the research in nanotechnology for 5 years. The Argentinean government, as well as the Brazilian, are trying to regulate all nanotechnology-related research by controlling budgets and by implementing supervisory procedures. But neither Argentina nor Brazil have set up discussion panels to examine the political, social and economic implications of the use of nanotechnology. In both countries, the exchange of ideas about the use of nanotechnology can only be associated with the idea of becoming more competitive (Foladori, 2006a).

The Mexican case is somewhat different from the Argentinean and the Brazilian. There is no specific plan nor national program linked to nanotechnology in México, even though nanotechnology is considered a strategic sector for development, as identified in 2002 in the *Special Program on Science and Technology 2001-2006* (Foladori & Zayago, 2007).

In the three countries, nanotechnology development has been supported by their governments. In fact, in order to develop the Millennium Scientific Initiatives (MSI), these governments had to disburse more money than the contribution made by the World Bank. The spirit behind these schemes is to support the development of leading technologies. That objective, leading toward greater competitiveness, justified the creation of enclaves of excellence. The argument is explicit in the case of Argentina and Brazil and very easy to discern from many documents issued by the Mexican government. In short, it is clear that the MSI are key elements of a policy directed to support the concept of knowledge economies through the establishment of Centers of Excellence.

The Millennium Initiatives of the World Bank gave initial, but very reduced, funding to the development of nanotechnologies in Latin America; particularly in Chile, Mexico and Brazil. Even though the MSIs pursued nanotechnology research in a very narrow way, they were some of the first to do so, thus they formed the core for encouraging nanotechnology research and development in those countries. In Chile, for instance, the programme began in 1999 and in Mexico in 2001. Ironically in these two countries there is still no programme or national fund to develop nanotechnology research. In Brazil, funding began in 2001, before the formation of the national networks of nanotechnology and three years before the elaboration of the Nanoscience and Nanotechnology Programme.

Although the spirit behind the MSI was to create research centers or institutes able to compete with their counterparts in developed countries, in practice they were under-funded. In Chile, for instance, the World Bank approved a credit of 5 million dollars for the first stage to be conducted in two and a half years, was added to the 10 million dollars approved by the Chilean government (ICM, n/d a). On average, most projects had a budget of 290,000 dollars for three years, with the possibility of just one renewal, which in the long term weakened the feasibility of the projects (Angel, 2003). In Brazil, during the first stage, the MSI received approximately 36 million dollars in funding for three years with an average of 2.1 million dollars per project (consisting of the formation of networks). The second stage was completely financed by the Brazilian government. In the Brazilian case, since the subjects were entire networks, the number of researchers and institutions involved was very large. The Nanosciences Institute, headquartered at the Federal University of Minas Gerais (UFMG) included 13 institutions and more than 60 Ph D researchers in almost 17 research projects (MCT-CNPQ, 2002).

The following chart shows the nanotechnology-related areas that were approved for these countries.

Millennium Science Initiatives in Latin America 1999 -2005[6]				
Country	**Start date**	**# of institutes/ nuclei/ networks funded**	**Nanotechnology Institutes or Nuclei founded by MSI**	**Host Institution**
Chile	**1999**	**3 Institutes 5 Nuclei**	**• Nuclei Condensed Matter Physics**	**• U Técnica Federico Santa María**
Chile	**2001**	**5 Nuclei**	**–**	**–**
Chile	**2002**	**3 Institutes**	**• Applied Quantum Mechanics and Computational Chemistry**	**• U Andrés Bello**
Chile	**2003**	**3 Institutes 8 Nuclei**	**• Nuclei Condensed Matter Physics (renewed)**	**• U Técnica Federico Santa María**
Chile	**2004**	**3 Institutes 12 Nuclei**	**–**	**–**
Chile	**2005**	**3 Institutes 15 Nuclei**	**–**	**–**
Chile (funds raised through "royalty fees:)	**2006**	**5 Institutes 17 Nuclei**		
				Contd...

[6] The information on Nanotechnology is approximate. The criteria employed was keyword indicators in the title or project description (nanotechnology, nanosciences, nanoscopic, nanostructured, nanocapsules).

Contd...

México	2001	4 Institutes	• Physiochemical Studies of novel Nanostructured Materials	• UASLP
Venezuela	2001 (closed)	3 Institutes; 8 Nuclei	–	–
Brazil (World Bank)	2001 - 2004	17 Networks	• Millenium Institute of Complex Materials • Nanosciences Institute UFMG • Research Network on System-on-a-Chip, Microsystems and Nanoelectronics	• UNICAMP • UFMG • UNICAMP

Source: ICM, 2006; CNPQ, 2005; MSI, 2005; & World Bank, 2005; ICM n/d.a; ICM n/d. b).

The goal for the MSIs was to link these research centers with industry, in public-private partnerships, to maintain money inflows into research, once the external funding expired. However, having the private sector finance scientific research is not standard operating procedure in Latin America, where most research is conducted in universities and public centers. In Brazil, more than 80% of the research is conducted in public centers. In some cases, the public-private partnerships for research are established, but this is not the general trend. On the few occasions when the public-private partnerships for research came into being, the private interest determined the research agenda.

How, then, can the private sector respond according to the national development interest, when research is motivated by profits and oriented toward the political economy of international competitiveness? That is a question seemingly without an answer within the nanotechnology development plans in Latin America (Foladori & Zayago, 2007). But, from the business perspective, the question can be answered with three arguments: research should be conducted according to the private interest, because this is the sector where the scientists are likely to work; when consumers buy a product they are seen to be confirming that the interest of the enterprise and the public is the same; and the amount and distribution of researchers required by the research centers of excellence will be regulated by the market itself according to the demand, so no shortage or surplus is expected. Therefore, the centers of excellence should not be seen as a process for creating elites, rather as the proper equilibrium between researchers and their demand within the market. This argument is charged with ambiguity because in reality inequality has increased in the last decades in the entire world, which questions its validity as the basis for development. The Mexican case is a good example, from the mid-1980s to mid-1990s, competitiveness significantly increased, but so did inequality, with the Gini coefficient going from 0.49 to 0.55 (Delgado Wise & Invernizzi, 2002).

Another problem without an apparent solution is sustainability from the point of view of the training process. The MSIs are an illustrative example of the knowledge enclaves, which the Chilean and Mexican experiences demonstrate. In several of these countries postgraduate programs of excellence are brought together with a deficient basic education and under-funded high school structures, where both are increasingly becoming privatized. In Mexico, only 19% of people of university age were enrolled; the figure for secondary education reached 57% in 2000 (Delgado Wise & Invernizzi, 2002). In addition, postgraduate programs of excellence are objects for external pressure since they are expected to produce a certain number of graduates, which then can jeopardize their quality (Guzmán del Próo, 2006). However, the most important question is whether these centers would be able to stop or even reverse brain drain, as it is hoped. This is a matter of concern, as research conditions in the centers of excellence in LDCs will never be of the same quality than their better-supported counterparts in developed countries. The latter is a consequence of the research agenda that prevails in developed countries where private interests rule. Consequently, the socio-political significance for researchers from LDCs, according to national development agendas, is lost. In fact, the private-public alliances encourage migration because such partnerships are established with transnational corporations and universities overseas, mainly in developed countries. This dynamic therefore facilitates personal relations and the insertion of these scientists in other contexts. Even thought the aim is to foster international competitiveness to encourage the process of economic growth, there is the problem of training researchers whose interests may not have anything to do with the national development agenda, understanding the latter as the reduction of poverty and inequality.

Conclusions

Development, in the economic growth-as-development perspective of the World Bank is unlikely to be achieved under the Milennium Scientific Initiative effort in Latin America. There remains too large a gap between the efforts of LDCs and the production of knowledge and corresponding patent production in countries which have better developed and funded research centres. This is particularly important in the essential foundation patents upon which future nanotechnology development will be based. That head start by the developed countries, combined with LDCs challenges in infrastructure and workers with the appropriate skill base to support an emerging nanotechnology industry, appears nearly insurmountable over the long term. Economic growth may well occur and appear in the national accounts as a boost to GDP, but the benefits will accrue to the south-north joint business partnerships, reproducing Latin America's long experience as the *labour* in service to multinational corporations and northern governments' *brains*.

From our broader development definition, one focused upon a decrease in poverty and an increase in equality, the repercussions of nanotechnology under the framework of the knowledge enclaves seems to be negative. Nanotechnology as an area for state investment or facilitation is not outlined within any national development plan beyond *economic growth* via an increase in competitiveness Those proponents of nanotechnology as a solution to the symptoms of underdevelopment in Latin America water quality, crop failure, medicines see the technology but not the structure under which that technology comes into play. Nanotechnology will create amazing new products and processes that may well have an impact upon our lives in a profound way but the reality is that LDCs will not own much of that technology: the structural challenges to ending poverty and inequality will remain, as the South works in service to the North, licensing patented technology from companies and governments from outside the region, while their capacity to create made-in-Latin America nanotechnology will remain circumscribed by insufficient planning, infrastructural deficiencies and a nanotechnology agenda set more by northern research partners than local needs.

(Dr. Guillermo Foladori, Professor, Doctoral Programme in Development Studies. UAZ, México. Member of the Latin American Nanotech & Society Network (ReLANS). He can be reached at fola@estudiosdeldesarrollo.net.

Mark Rushton, PhD. Candidate, Doctoral Programme in Development Studies, UAZ, México. He can be reached at mrushton@mac.com

Edgar Zayago Lau, PhD. Candidate, Doctoral Programme in Development Studies, UAZ. Member of the Latin American Nanotech & Society Network (ReLANS). He can be reached at México.edzlau@yahoo.com.)

References

Angel V., Eliette (2003). Iniciativa Científica Milenio. Química y neurociencia no tienen fondos. El Mercurio, September 29, 2003. *http://www.ceo.cl/609/printer-48524.html* [Last visited May 22, 2007].

Ahooja-Patel, Krishna (1982). Another Development with Women. *Development Dialogue*, 1(2).

Baran, Paul (1957). *The Political Economy of Growth.* New York: Monthly Review Press.

Brahic, Catherine (2005a). Developing world 'needs nanotech network'. *SciDev.Net* February 11, 2005. *http://www.scidev.net/news/index.cfm?fuseaction=printarticle&itemid=1923&lan guage=1* [Last visited July 27, 2006].

Brahic, Catherine (2005b). Nanotech revolution needs business know-how. *SciDev.Net* February 18, 2005. *http://www.scidev.net/News/index.cfm?fuseaction= readnews&itemid= 1938&lang uage=1* [Last visited July 27, 2006].

Brahic, Catherine and David Dickson (2005). "Helping the poor: the real challenge of nanotech". *SciDev.Net* February 21, 2005. *http://www.scidev.net/content/editorials/eng/ helping-the-poor-the-real-challenge-of-nanotech.cfm* [Last visited July 27, 2006].

Butler, Declan (1998). World Bank calls for a fairer deal on patents and knowledge. *Nature*, 395, 529.

Chambers, Robert and Conway Gordon.

(1998). "Sustainable Rural Livelihoods: Some Working Definitions". *Development*, 41 (3). September.

(1995). *Poverty and Livelihoods: Whose Reality Counts?* Brighton: IDS, University of Sussex.

Chambers, Robert (1987). *Sustainable Rural Livelihoods: A Strategy for People, Environment and Development.* Brighton: IDS University of Sussex.

CNPQ (2005). Institutos do Milênio. *http://www.cnpq.br/programasespeciais/milenio/projetos/2005/05.htm* [Last visited May 20, 2007].

Cypher, James (2005). "The Political Economy of the Chilean State in the Neoliberal Era", *Canadian Journal of Development Studies*, XXVI, 4, 763-779.

Delgado Wise, Raúl & Invernizzi, Noela (2002). México y Corea del Sur: Claroscuros del crecimiento exportador en el contexto del globalismo neoliberal. *Aportes, Revista Mexicana de Estudios sobre la Cuenca del Pacífico*, II, 2, 4, 63-86.

Dickson, David (2005). "G8 leaders give indirect boost for science in Africa". *SciDevNet*, September 3. *http://www.scidev.net/news/index.cfm?fuseaction=printarticle&itemid=2549&lan guage=1* [Last visited September 1, 2006].

DORCH (Diario Oficial de la República de Chile) (1999). Decreto No. 151. Julio 27, 1999. P.4 (7308). Ministerio de Planificación y Cooperación. Crea Comisión Nacional de Iniciativas Científicas para el Milenio.

ETC, Group (2003). "From Genomes to Atoms: The Big Down/ Atomtech: Technologies Converging at the Nano-scale". *http://www.etcgroup.org/upload/publication/171/01/thebigdown.pdf*[Last visited may 24, 2007].

Foladori, Guillermo. (2006). *Nanotechnology in Latin America at the Crossroads Nanotechnology Law & Business Journal*, 3(2), 205-216.

Foladori, Guillermo & Zayago, Edgar. (2007-forthcoming). Tracking Nanotechnology in Mexico. *Nanotechnology Law & Business*, September issue.

Friedman, Milton.

(1980). *Free to Choose: A personal statement.* New York: Harcourt Brace.

(1962). *Capitalism and Freedom*. Chicago: University of Chicago Press.

Guzmán del Próo, Sergio A. (2006). Los posgrados y la formación de recursos humanos. La Jornada. México D.F. *http://www.jornada.unam.mx/2006/07/26/a03a1cie.php* [Last visited July 26, 2006].

Hassan, Mohamed (2005). "Nanotechnology: Small things and big changes in the developing world". *Science*, 309, 5731, 65-66.

Hayek Friedrich A. (1944). *The Road to Serfdom: with foreword by John Chamberlain.* Publisher: Chicago: University of Chicago Press.

ICM. (2006). *Iniciativa Científica Milenio. Memoria Trienal 2003-2005. www.mideplan.cl/milenio/?q=node/*113 Last visited May 12, 2007.

ICM. (n/d a). *Iniciativa Científica Milenio. Memoria Bianual 1999-2000.* Santiago: MIDEPLAN. *http://www.mideplan.cl/milenio/?q=node/34* [Last visited May 23, 2007].

ICM. (n/d b). *Iniciativa Científica Milenio. Memoria Bianual 2001-2002.* Santiago: MIDEPLAN. *http://www.mideplan.cl/milenio/?q=node/35* [Last visited May 23, 2007].

Invernizzi, Noela, and Guillermo Foladori. (2005). "Nanotechnology and the Developing World: Will Nanotechnology Overcome Poverty or Widen Disparities?" *Nanotechnology Law & Business Journal.* Vol. 2. Iss. 3. Article 11.

Invernizzi, Noela; Foladori, Guillermo & Maclurcan, Donald (2007). The role of nanotechnologies in development and poverty alleviation: a matter of controversy. (*Draft. Offered under request*).

Juma, C. & Yee-Cheong, L. (Eds.). (2005). *Innovation: applying knowledge in development.* UN Millennium Project. Task Force on Science, Technology, and Innovation. London, Sterling, Va.: Earthscan. *http://www.unmillenniumproject.org/documents/Science-complete.pdf* [Last visited, 29 May, 2007].

Macilwain, Colin (1998). World Bank backs Third World centres of excellence plan. *Nature,* 396, 711, 24-31.

Masood, Ehsan (1999). El Banco Mundial invierte en una base científica global. *Nature,* 397, 6-7.

MCT (Ministério da Ciência e Tecnologia). (2004a) O Programa de Nanotecnologia. *http://www.mct.gov.br/Temas/Nano/programanano.htm* [Last visited January 05, 2006].

MCT (Ministério da Ciência e Tecnologia). (2004b). Portaria MCT nº 614, de 1º.12.2004. *www.mct.gov.br/legis/portarias/614_2004.htm* [Last visited October 10, 2005].

MCT-CNPq (2002). A Iniciativa Brasileira em Ciência e Tecnologia. *Parcerias Estratégicas,* (2004), 18, 105-135.

MSI (Millennium Science Initiative). (2005) Current and Planned Initiatives. *http://www.msi-sig.org/msi/current.html* [Last visited May 12, 2007].

Nature. (1998). Urgent thinking required about development. *Nature,* 395, 6702, 527.

OECD (Organisation for Economic Co-operation and Development) (2005), *Main Science and Technology Indicators 2005-2.*

Prebisch, Raul

— (1950). The Economic Development of Latin America and its Principal Problems. Reprinted in *Economic Bulletin for Latin America,* Vol. 7, No. 1, 1962, 1-22.

— (1984). Five Stages in my Thinking on Development. In Meier and Seers (eds.) Pioneers in Development. New York: Oxford University Press Rostow, W.W. (1960). *The Stages of Economic Growth: A Non-Communist Manifesto.* Cambridge: Cambridge University Press.

Salamanca-Buentello, F.; Persad, D. L.; Court, E. B.; Martin, D. K.; Daar, A. S.; Singer, P. (2005). Nanotechnology and the Developing World. *PLoS Medicine*, 2 (5), p/?. *http://medicine.plosjournals.org/perlserv/?request=get-document&doi=10.1371/journal.pmed.0020097* [Last visited May 22, 2007].

Sen, Amartya (1988). "The Concept of Development. In Chenery and Srinivasan" (eds.) *Handbook of Development Economics,* Vol. I. Amsterdan: North Holland.

Streeten, Paul (1994). *Strategies for Human Development: Global Poverty and Unemployment.* Copenhagen: Munksgaard International Publishers.

Tanthapanichakoon, W. (2005). "An Overview of Nanotechnology in Thailand", *KONA*, 23, 64-68.

Tun Razak, N. (2005). Speech by the Honourable Dato' Sri Mohd Najib Bin Tun Abdul Razak, the Deputy Prime Minister of Malaysia. *Malaysian Nanotechnology Forum*. Johor: Universiti Teknologi Malaysia.

UNDP. (2006). *Human Development Report* 2006. *http://hdr.undp.org/hdr2006/statistics/indicators/156.html* [Last accessed 23 May 2007].

UNDP. (1990) *Human Development Report 1990.* New York: Oxford University Press.

Valdéz, Juan Gabriel (1989). La escuela de Chicago: Operacion Chile. Buenos Aires: Grupo Editorial Zeta, S.A.

World Trade Organization. (1994). TRIPS, Annex 1C *http://www.wto.org/english/docs_e/legal_e/27-trips.pdf* [Last visited May 22, 2007].

World Bank. (2006). *World Development Indicators 2006.* CD-ROM. Washington, D.C.

World Bank. (2005). Implementation Completion Report on a Loan in the Amount of USD $5.0 Millions to the Bolivarian Republic of Venezuela for the Millennium Science Initiative Project (scl-45720) *http://www-wds.worldbank.org/external/default/WDSContentServer/WDSP/IB/2005/06/27/00 0012009_20050627102527/Rendered/PDF/317310rev.pdf* [Last visited May 12, 2007].

World Bank. (1999). *World Development Report 1998/99.* New York: United Nations.

11

Nanotechnology
New Vistas for US

D Gayatri

According to national Science Foundation (NSF) of the US, the global market for nano product was estimated at around $1 trillion by 2015 and about $7.6 billion by 2003. All the possible benefits that the technology could bring for the economy were encouraging the state and federal government to invest in research activities and infrastructure for building nanotech hubs. With the Government support many regional nanohubs emerged. In addition, venture capitalists also seemed to be interested in this field and invested around 2225 million in the first quarter of 2003. The nanotech bill signed by President George Bush Jr., is a vital catalyst for the development and growth of what will become a $ 1trillion piece of the global economy. This bill will also allow the US to continue to strive for global leadership in this highly competitive nanotechnology marketplace. The bill according to analysts was a model of how the companies, Government and the universities could work together towards the growth of the economy.

Source: The Icfai Business School Case development Centre, Hyderabad.

"A cluster is a geographically proximate group of companies and associated institutions in a particular field, linked by commonalities and complementarities."

– Michael E Porter.[1]

"Nanotechnology may lay the groundwork for a new Industrial Revolution."

– Scientific American, September 2001.

"The convergence of nanotechnology with information technology, biology and social sciences will reinvigorate discoveries and innovation in many areas of the economy."

– George W Bush, President of the United States.[2]

Introduction

Nanotechnology refers to the concept of controlling the properties of material at the atomic level. This technology aims at making things smaller and smaller, which could be measured in nano scales and whose properties like color, hardness etc., could be controlled.[3] The idea was to actually make things atom by atom. The past innovations of Silicon Valley were results of adverse conditions like World War II leading the investments in defense and foreign competition demanding expenses on semiconductor industry. Incidences like the September 11 were considered as promoting investments in Nanotechnology indirectly because Government identified the importance of Nanoscience in fighting terrorism.[4] Nanotechnology could be used to make powerful weapons and sensors to detect bio-terror scares like anthrax easily. Nanotechnology was also expected to reduce the country's energy dependence in future.

According to National Science Foundation (NSF) of the US, the global market for nano products was estimated at around $1 trillion by 2015 and about $7.6 billion by 2003. The industry was expected to offer around seven million jobs during the next 10-15 years.[5] Among the Nanoproducts, the Nanomaterials, which were the silver crystals and other Nanoparticles used in photographic films, accounted for the highest share (97.5%) of the market.[6] There were around 1,500 Nanotechnology start-ups around the world, out of which around 1000 were in the US. Most of the companies operated in the sector of Nanodevices followed by Nanoinstruments and Nanobiotechnology (Annexure I).

All the possible benefits that the technology could bring for the economy were encouraging the state and federal government to invest in research activities and infrastructure for building Nanotech hubs. The Nanotechnology industry was allocated an amount of $3.7 billion by the US congress in 2003. This initiative was a part of the National Research and Development Act released by the Government.[7] With the government support, many regional nanohubs emerged in the states of Texas and New York among others. In addition, Venture Capitalists also seemed to be interested in this field and invested around $225 million in the first quarter of 2003.[8]

Background

The applications of Nanotechnology ranged from imaginable ones of making things stronger and lighter by working precisely on each atom at a time and eliminating waste, to unimaginable applications such as using carbon Nanotubes that were 100 times stronger than steel and only one sixth of their weight, to build space elevators. The idea of making small things started in 1959, when Richard Feynman, thought of writing the whole Encyclopedia Britannica on a pinhead. The idea was to work on the products at a molecular level to make things smaller. In his famous lecture "There is plenty of room at the bottom", he defined the concept of molecular manufacturing and miniaturization of instruments, which could be used to measure the properties of the nano structures.

The Silicon Valley was an outcome of unplanned events, which started at the Stanford University. Federick Terman, father of Silicon valley, was the first person to think of clustering different participants in the 1950s, that ultimately led to a successful implementation of technological research in the then growing electronics industry.[9] In the case of Nanotechnology, it was the Rice University, whose professors discovered a new form of carbon in 1985 that laid the foundation for the new era of nanoscale manufacturing. Again in 1986, researcher and author Eric Drexler's seminar work on nanotechnology added to the growing popularity of the concept.

Nanotechnology: Some Applications

The Nanotech era started with the Scanning Tunneling Microscopes developed by IBM in 1981 and the Atomic Force Microscopes that were released in the market in 1989 that took atomic scale images of metal surfaces which in turn were used for detecting surface defects and measuring the size and number of molecules. Using these microscopes scientists could see the atomic landscapes of materials. People could actually move atoms with the precision they obtained through these microscopes. In 1990, IBM made a remarkable achievement by making the smallest logo of the world. It assembled 35 Xenon atoms at a minute scale to make its logo that spanned only 3 nanometers.[10]

Exhibit I: Smallest Logo Developed by IBM by Moving Atoms using STM

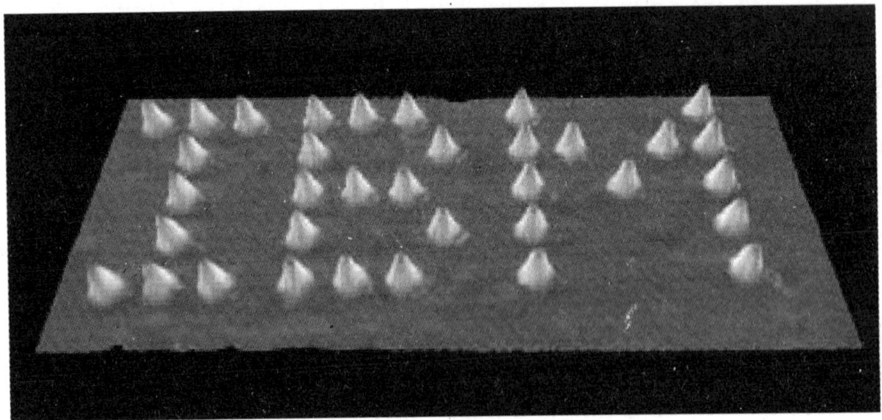

Nanotechnology applications spanned various fields like medicine, molecular manufacturing and space research programs, using Nanotechnology involved monitoring and controlling the manufacturing process at the molecular level. While the major benefits of the technology were nascent, some of the products utilizing this technology were already in the market (Annexure II). The first company to use Nanotechnology was the Zyvex Corporation, a Texas based firm. It was started in 1997 by John Von Her II. Their mission was "to be the leading worldwide supplier of tools, products, and services that enable adaptable, affordable, and molecularly precise manufacturing". The company spent around $100,000 on market research to find out the feasible applications for the new technology and identified areas like aerospace, defense, health care, and electronics & telecommunications. Zyvex was later awarded a five year contract by the NIST (National Institute of Standards and Technology) for manufacturing Nano-scale, low-cost assemblers for which it obtained a fund of $25 million.[11] But, The most popular and visible applications were the stain resistant fabrics developed by Nano-Tex called the Nano care, a coating for fabrics that repelled against stains. Tiny hair like fibres were embedded into fabrics, which acted against stains sticking to clothes. Levi was using this technology for Jeans and performance khakis.

Contd...

Contd...

Other products were the transparent sunscreen lotion of Advanced Power Technologies, NanoDefend and NanoGreen that could be used to decontaminate clothes and skin, VS Nanotube power rackets developed by a French manufacturer Babolat, which used carbon Nanotubes supplied by France's Nanoledge to make the rackets lighter and stronger, and the Double core tennis balls manufactured by Inmat whose Nanoclay materials made the balls last longer than before.[12]

Among the future applications were the advanced drug delivery systems, medical diagnostic tools, and solar cells that could make power inexpensive. The displays of laptops, cell phones could be made out of Nanostructured polymers that were formally called the OLEDs (Organic Light Emitting Diodes). These displays were less expensive, brighter and less power consuming compared to other normal diodes. The Dow chemical came out with Nanoparticle-reinforced polymers that could replace metallic components of automobiles making them much lighter and less fuel consuming. It was expected that this technology would reduce the gasoline consumption of the country by 1.5 billion litres by 2003.[13]

The technology could also be used for the development of super quality chips in future. A new form of carbon, discovered by the Smalley group of the Rice University, was expected to replace silicon chips in the next 15 years. Dupont, a chemicals company, sold its businesses of nylon, polyester and Lycra to take up Nanotechnological research to develop an integrated science using the insights of molecular biology. Companies were keen on exploring the technological benefits as fast as they could. Nanosys, a California based company was using Nanotechnology for all its products ranging from solar cells to computers and already had raised around US$ 39 million by 2003.

Source: www.almaden.ibm.com

Emerging Nano-Hubs and Government's Role

Commenting on the future of the technology, Richard Smalley, Nobel Laureate, said, "We are about to be able to build things that work on the smallest possible length scales. It is in our nation's best interest to move boldly into this new field."[14] To repeat the success of Silicon Valley, one of the most successful clusters in the world, Nanotechnology hub had to also replicate the competitive advantage the clusters could deliver for any business. Clusters promoted competition through collaboration among companies, universities and funding institutes. In addition, clusters helped companies attract venture capitalists and qualified personnel for research. According to management guru, Michael E Porter, it was believed that a competitive advantage could be gained by companies of a country over their counterparts in other countries through clusters.[15] This was possible if the government invested in research that could be shared by all the companies. The US government led by the George Bush administration was keen on developing more of such clusters, which could contribute to the growth in the economy. Most of the successful clusters in America showed the government support to be the prime reason for their success. But, the interdisciplinary nature of the science demanded clusters to be located in centers that were accessible to the customers and the venture capitalists, close to the schools and academia and not out of reach of companies who could take advantage of the clusters, the analysts felt.[16] In the late 1990s, the semiconductor industry and the universities jointly proposed for federal funding.[17] In 1997, the government invested around $116 million on Nanoscience research and engineering.[18]

The government was also established funding networks to help start-ups. In the initial years, Venture capitalists, though were keen on investing, were relatively unsure of the success. So, the government's financial network called the Nanobusiness Angel Network appointed financial experts to evaluate companies that were aspiring for research and were looking out for funds. This network aimed at catalyzing the research activities in the field by forming hubs, and bringing together researchers, investors, enterprises and government bodies.[19]

Identifying the importance of the new technology for the future of the country, the government established the National Nanotechnology Initiative (NNI) in 2000. The NNI coordinated the activities of participants in the field of Nanotechnology. Then the president of America, Bill Clinton announced an additional investment of $227 million in Nanotechnology research. This initiative funded the following agencies, National Science Foundation (NSF), the Department of Defense (DOD), the Department of Energy (DOE), National Institute of Health (NIH), National Aeronautics and Space Administration (NASA), and Department of Commerce's National Institute of Standards and Technology (NIST)[20] (Annexure III).

The NanoBusinessAlliance – a New York based industry group was established in October 2000, which focused on collaborating activities to facilitate the formation of hubs. It set up its first regional hubs in Washington and Denver in the beginning of 2001. It was planning to establish 15 new hubs by 2002 in Virginia, New Mexico, Boston and other centers. The goal was to provide a common platform for understanding, planning and implementing Nanotechnology businesses in the country. NanoBusinessAlliance was also at the same time identifying the nanotech products that could be commercialized within a small timeframe to encourage private funding. According to Edward Moan, director of product innovation, Deloitte & Touche, a consulting firm, "The hubs will provide nascent nanotech companies with the visibility necessary to attract world-class funding sources and to reach their potential customers." The goal was to educate people about the realistic impact the technology could bring for various industries.[21]

The major regions in the US that were contending for the title of Nanotech capital of the US through their regional hubs were California, Texas, New York and the Pennsylvania (Annexure IV). Pennsylvania, whose investment authority funded around $10.8 million to the state's Nanotechnology Initiative in 2000, was emerging as a possible future hub. The state had the largest number of pharmaceutical companies than any other state and so the possible applications of the Nanotechnology in the medical field were large. On the other hand Virginia was concentrating on research than commercialization because of the proximity to government funds from National Science Foundation and US Department of Defense. Boston received huge funds and had institutions like MIT and Harvard to facilitate quality research. IBM in New York was steadily investing in research. Lux Capital, the Nanotech venture capital firm based in New York invested big amounts. Each area enjoyed an advantage over the other.[22]

The next in the line was California with the intellectual back-up from Stanford, Berkeley, Caltech and the University of California. In addition, it had a proximity to the large pool of qualified engineers and scientist from the Silicon Valley. It also had a huge source of Venture Capital funds. In 2002, it received around $450 million from the state government for research, which was around 50% of total investment of funds in the US.

Texas was also a major aspirant. Zynex contributed around $2.5 million for Nanotechnology research in the University of Texas at Dallas. It was also behind the Texas Nanotechnology Initiative, which aimed at collaboration of researchers, government, and Venture Capitalists. Texas had an advantage of developing faster than any other state because of its tax policies that charged no personal or corporate tax. In addition, proposals were under consideration to totally nullify tax on research activities.[23] Texas based A&M University was awarded a contract by NASA to develop the next generation Nanoproducts for aerospace by working with six other top Universities in Texas.[24] This program was led by the Boeing Company, which was already working towards making "super intelligent planes" which would automatically recover from breakdowns. The Nanotechnology was also expected to make airplanes stronger and lighter in the next 20 to 50 years when the full benefits of the technology would be actually identified.[25] The University of Texas spent around $500,000 towards nanotechnology research. In 2002, a non-profit foundation, National Foundation of Texas, was established, which funded around $1 million annually to Universities over the next five years.[26]

Venture Capital

Private funding formed only a small part of the research funding. Of the total funds that went into Nanotechnology research, only 4% was contributed by Venture Capital (VC) industry, with 29% contribution from the federal government and 34% from the companies.[27] The federal government was playing a major role.

But, after the Nanotechnology Initiative by the government, the venture capitalists started showing interest (Annexure V). By 2002, the venture capitalists' investment into Nanotechnology had reached around $880.3 million, which accounted for around 4.2 % of the total venture capital in the US. The lab to market time for these products was around 10-15 years and the commercial benefits of a particular research were unimaginable till it actually was completed. But, slowly the scenario was changing, Venture Capital firms like the LUX capital of New York and the Draper Fisher Jurveston of Redwood city were hiring staff for investigating the possible benefits the research could bring to the firms.[28]

By 2002, around 20 venture capital firms were eager to invest in research. But it was felt that the successful commercialization of Nanoscience in future would be possible only if various parties joined hands. According to analysts, the formation of clusters in more number was essential for the US to grow on the new technology, with other countries such as Japan and Europe spending heavily in the field (Annexure VI). At the same time, the interest shown by the major companies such as IBM, Hewlett Packard, Motorola in the Nanotechnology was encouraging private investors.[29] Around 700 public and private companies were working on this technology in the US by the end of 2003. Nanosys Inc., the leading Nanotechnology Company raised an amount of $30 million in spite of doubts raised about the success of research in the field. It managed to fund its research from major venture capitalists including Lux Capital and Harris & Harris. The venture capitalist investment was the highest in California followed by New York, Pennsylvania and Texas.

Nanotechnology – Is the Dream Big?

The President of America, George Bush Jr. signed the Nanotechnology Research & Development Act in November 2003, which allotted $3.7 billion for research and implementation of projects.[30] The bill also laid the foundation for the coordination between the industry and the academia. It set up a joint advisory board with representatives from academia and industry that would coordinate short-term to long-term research programs. The act also put forward an action plan for monitoring the progress of research programs that would be undertaken by the National Science Foundation, the department of energy, the department of commerce, NASA and the environmental protection agency. A panel for assisting the President in the technology issues was also proposed and approved by the bill.[31] Training of professionals in this field was also a part of the act. The bill also set up an advisory committee that administered the commercialization of the research. According to F Mark Modezelewski, Executive Director, NanoBusinessAlliance, "This nanotech bill is a vital catalyst for the development and growth of what will become a $1 trillion piece of the global economy. This bill will also allow the US to continue to strive for global leadership in this highly competitive nanotechnology marketplace." The bill, according to analysts was a model of how the companies, government and the universities could work together towards the growth of the economy.[32]

In 2003, Rice University and IBM went into an agreement according to which IBM would provide high end computing to the Nanotechnology research activities that would be undertaken by the university. The Rice center for Biological and Environmental Nanotechnology would receive the super computing system of IBM, the IBM eServer p60. This system would help the institute to solve some of the complex mathematical problems of molecular structures.[33] Around 13 universities in America joined hands to develop a common research platform for Nanoscience. The Cornell University would lead the operations under the National Nanotechnology Infrastructure Network (NNIN), which was expected to start in 2004. According to Lawrence Goldberg, senior engineering advisor, National Science Foundation (NSF), an investment of $700 million would be kept in building this network. The NNIN was to provide the necessary infrastructure for research, including the professional help and latest lab tools to the participating universities. In addition, the network would also inspect the social and ethical impacts of the technology and ways to overcome them.[34]

In 2004, Dupont, which was steadily investing in research in this field, donated its patents to the University of California-Davis. The process of generating electronic beams using Nanotechnology could be used to generate high intensity lighting, more accurate medical devices, and nanoscopic X-rays. The company was enthusiastic about sharing its research with universities, which could be used by them for further research.[35]

Moreover, companies were also collaborating with each other to improve the quality of their research activities. The Combi Matrix Group and the Cyrano sciences were collaborating to explore the chemical side applications of nanotechnology. While the former was a company exclusively manufacturing Nanomaterials, the latter was involved in manufacturing of products that digitized human senses like the speech recognition devices and digital cameras.[36]

Analysts felt that the partnerships with international counterparts would be essential to monitor the activities of other countries.[37] It was also felt that the Bush administration and the congress should work together to double their spending on research and that America's leadership in the 21st century would depend on how well it invested in these projects which had long-term benefits.[38] As John Marburger, Director of Office of Science and Technology policy put it, "Investments in nanoscale science and technology research and development are essential in achieving the President's top three priorities: winning the war on terrorism, securing the homeland and strengthening the economy."[39]

(D Gayatri, Icfai Business School Case Development Centre, Hyderabad. She can be reached at dgayarri@icfaiorg, gayatri@rediffmail.com.)

Endnotes

1 "Clusters", *www.competitiveness.com.*

2 "What is nanotechnology", *http://newsletters.forbes.com/nanotech.*

3 A nanometer is one billionth part of a meter.

4 "Next silicon valley: riding the waves of innovation white paper", *www.jointventure.org,* December 2001.

5 "Council calls for expanding Nanotechnology", *www.bizjournals.com,* January 24th 2002.

6 Rajan, Malika, "Global nanotechnology market to reach $29 billion by 2008", *www.bccresearch.com,* February 4th 2004.

7 "Nanotechnology Research and Development Act becomes law", *www.sensormag.com,* November 24th 2003.

8 Buff, Philipp, Hermann, "Venture Capital and investements", *www.swissnanotech.net,* August 2003.

9 Markale,Ganesh, "A modern history of silicon valley", *www.ciol.com,* March 7th 2000.

10 "Nanotechnology Shaping the world atom by atom", *http://www.wtec.org/loyola/nano/IWGN.Public.Brochure*

11 "Accelerating the nanotechnology revolution: Zyvex's Market Driven Strategies", *www.glocom.org,* February 10th 2003.

12 Seminar delivered by Philip.J.Bond, Undersecretary of commerce for technology, US department of commerce, at the NNI in Washington in 2003, *http://seclists.org,* April 8th 2003.

13 "Nanotechnology- prepared written statement and supplemental material of R.E.Smalley", Rice University, *www.house.gov,* June 22nd 1999.

14 Roco, C., Mihail, "A frontier for engineering", *www.memagazine.org*

15 Michael E. Porter coined the word Clusters in his book "*The Competitive Advantage of Nations*"

16 Stewart, Duncan, "Cluster- Busters", *National post, www.teracap.com,* November 8th 2003.

17 Leopold, George, "Legislation seeks clearer U.S nanotechnology plan", *www.eetimes.com,* October 18th 2002.

18 Roco, M.C, "Government nanotechnology funding: An international outlook", *www.nsf.gov.*

19 Mason, Jack, "Nanobusiness alliance launches angel network and regional hubs", *www.smalltimes.com,*

20 Terra, P., Richard, "National Nanotechnology Initiative in FY2001", *www.foresight.org.*

21 Burnell, R., Scott, "Business group spreads word on nano", *www.upi.com*, February 15th 2002.

22 Richardson, "Regions across the nation vie for Nanotech capital title", *www.smalltimes.com*, July 16th 2001.

23 "Nanotechnology foundation of Texas Inc.," *www.nanotechfoundation.org*, April 2003.

24 Beck, Dan & Beach, Ann, "Boeing led team wins contract to advance nuclear electric power for space", *www.boeing.com*, October 3rd 2002.

25 Swanson, Sandra, "Smarter, safer planes via nanotech", *www.informationweek.com*, December 17th 2001.

26 "Nanotechnology white paper", *www.nanotechfoundation.org*

27 Fried, Jane, "VCs are more talk than action in funding nano, expets say", *www.smalltimes.com*, July 26th 2002.

28 Ibid.

29 "Nanotechnology wins over mainstream venture capitalists", *www.redherring.com*, December 18th 2001.

30 Mark, Roy, "Bush signs Nanotechnology bill", *http://dc.internet.com*, December 4th 2003.

31 Mark, Roy, "Senate committee approves Nanotech R&D bill", *http://dc.internet.com*, June 19th 2003.

32 Arya,"Senate Passes Nanotechnology Bill; NanoBusiness Alliance Backed Legislation Will Help to Propel the Future $1 Trillion Global Nanotechnology Market", *www.nanoinvestorsnews.com*, November 19th 2003.

33 "Rice university announces NanoTech deal with IBM", *www.spacedaily.com*, January 31st 2003.

34 "National Science Board Approves Award for national nanotechnology infrastructure network", *www.nsf.gov*, December 22nd 2003.

35 "Dupont donates Nanotechnology patents", *www.nanoxchange.com*, February 3rd 2004.

36 "Nanotechnology companies collaborate", *www.reed-electronics.com*, January 12th 2004.

37 Leopold, George "Congress set to boost Nanotechnology funding", *www.techweb.com*, June 28th 1999.

38 Kalil, Thomas, "Cutting edge small technology deserves big US investment", *http://archives.seattletimes.nwsource.com*, June 28th 2001.

39 "White house reports touts nanotechnology", *www.eetimes.com*, October 17th 2003.

Annexure I: Number of Companies Operating in Various Nanotech Sectors in 2004

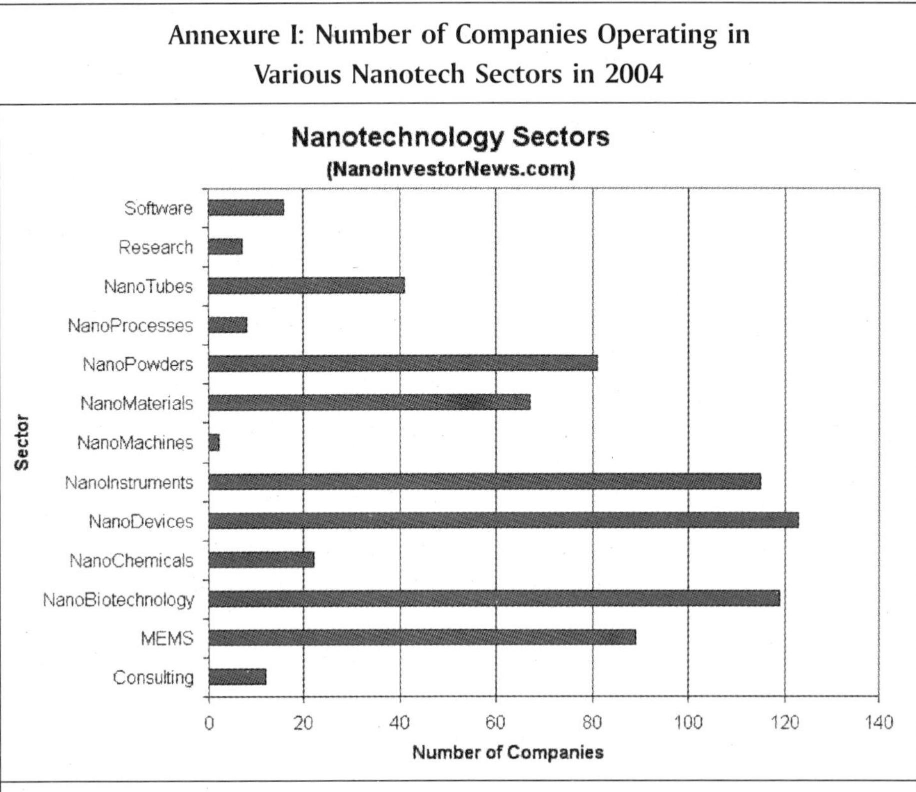

Source: www.nanoinvestornews.com

Annexure II: Top Ten Nanoproducts in 2003

- High Performance ski wax: Developed by German based company Nanogate. Nanowax is a polymer that makes ski and snowboard bases harder and faster gliding. It hardens when the temperatures drop, adjusting to temperature changes. These coatings last longer than usual waxes.
- Waterproof Ski Jacket: Franz Ziener GmbH & Co., a German based company. The jackets are water proof, windproof and last longer.
- Wrinkle and Stain repellant Threads: Nano care, developed by Nano-Tex, a California based company that made wrinkle resistant and stain repelling fabrics using them. Used by Lee jeans.
- Deep-Penetrating skin creams: Plenitude Revitalift antiwrinkle cream by L'Oreal Paris, that embeds vitamin A into polymer capsules, which act as sponges that hold the cream until they get absorbed into the skin.

Contd...

Contd...

- World's First digital Camera: Cameras from Kodak that use OLEDs that make the displays brighter. They don't require lighting.
- DVD and Books collection: DVDs and books on Nanoscience.
- Sunglasses: Sunglasses from Smith, uses nanofilm to produce protective, antireflective, ultrathin coating for glasses.
- Nanocrystalline sunscreens: Z-COTE, a sunscreen from BASF, uses high-purity nanocrystalline zinc oxide, which allows the sunscreen to go on clear. In addition the inorganic Z-COTE can't be absorbed by the skin and won't cause allergic reactions.
- High-Tech Tennis Rackets And Balls: Introduced by France based Babolat, Nanotube Power racket that are five times more harder than usual rackets. Balls from Wilson Double, which uses Air D-fense, nanocomposite product from Inmat New Jersey based nanocompany. These balls can be used for four weeks. Selected as the official balls for Davis cup.

Source: www.forbes.com

Annexure III: Government's Investment in Various Departments
Summary of Federal Nanotechnology Funding ($ in Millions)

Department/Agency	FY 2000	FY 2001	FY 2002	FY 2003
NSF	$97	$150	$199	$221
Defense	70	110	180	201
Energy	58	93	91	139
NASA	5	20	46	51
NTH	32	39	41	43
NIST	8	10	38	44
Total (incl'g other)	$270	$422	$604	$710

Source: National Science Foundation.

Annexure IV: Distribution of Nanotech Companies in Various Regions in the US in 2004

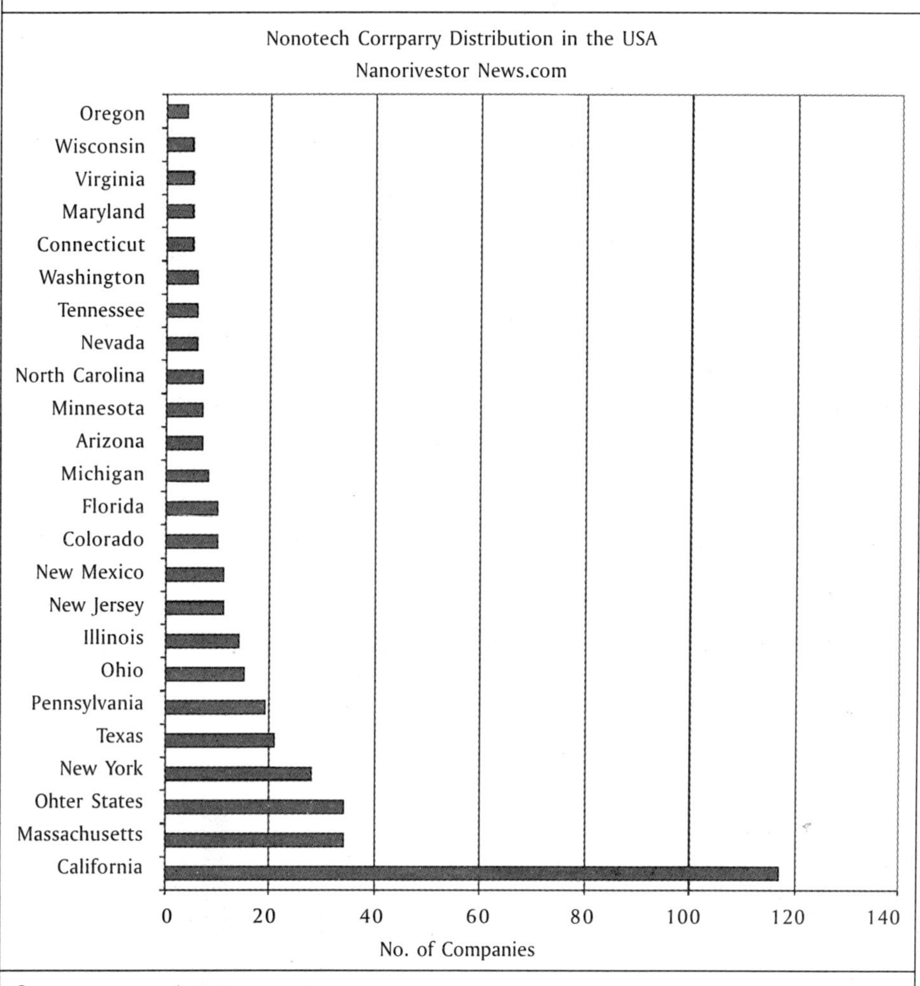

Source: www.nanoinvestornews.com

Annexure V: Government Funding and Venture Capitalist Funding for Nanotechnology in the US from 1999 to 2003

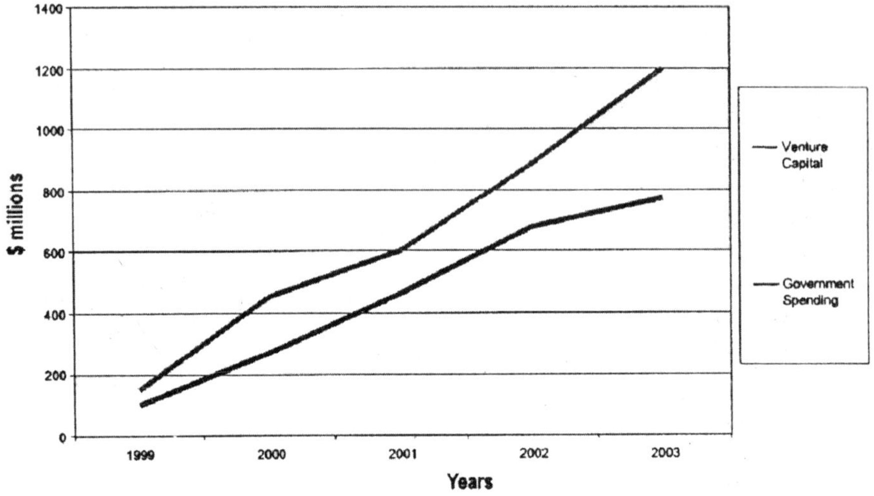

Source: www.osec.ch

Annexure VI: Spending on Nanotechnology in Europe, US and Japan

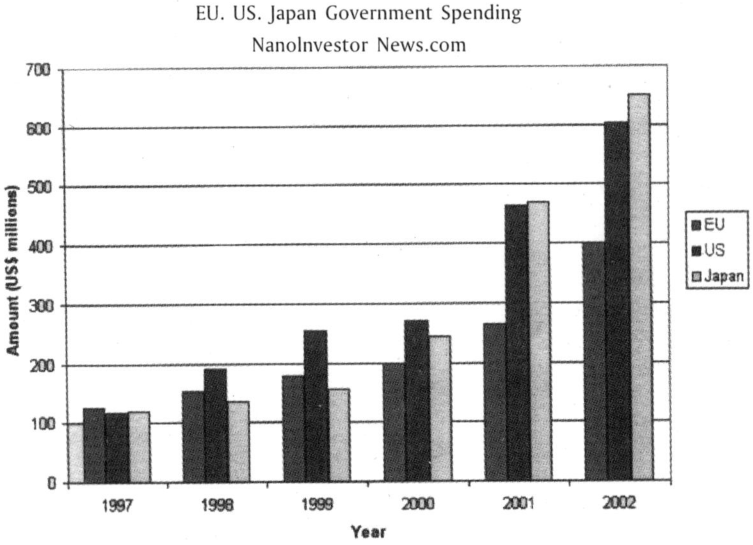

EU. US. Japan Government Spending

NanoInvestor News.com

Source: wwwnanoinvestornews.com

Developments in Nano Products

– Anil Varma

When giant consumer electronics from South Korea – Samsung – introduced home appliances which had coating of nano silver which is a compound that wipes out as well as stops the spreading of bacteria, the concept they were depending on was that of hygienic and clean living. The concept is to use nanotechnology so as to keep the food fresh, clothes free of germs and the air pure and clean.

Samsung has been very active in its engagement with nanotechnology and the company has been putting the technology into operation to items that are really significant on a large scale, namely; semiconductors, components of telecommunication and flash memory. The organization is associated at close quarters with the South Korean government with relation to the national nanotechnology agenda. Samsung is also working in cooperation with the Interuniversity Microelectronics Center which is one of the premier self-sufficient research institutes in the field of nano electronics, on various programs so as to establish in what manner can nanotechnology, assist in the progress of the future generation of portable devices of communication, namely in mobile phones, televisions and digital cameras. It is anticipated that the beneficial results of these programs will soon be visible in the market.

(Anil Varma, Consulting Editor, The Icfai Research Centre Pune. He can be reached at avarma@iupindia.org.)

12

Societal Implications of Nanoscience and Nanotechnology in Developing Countries

Birgit R Bürgi and T Pradeep

The historically unprecedented developments of nanoscience and nanotechnology, in view of their phenomenal expansion and growth, in conjunction with their convergence with information science and molecular biology, confront our society and natural environment with new challenges. Technological revolutions have shown that discoveries at the frontiers of science have the potential to pave the way for radically innovative and integrated approaches, providing new solutions for some of the most pressing problems. In order to enable decision-makers to respond to what is best for people at large, the societal implications of the newly emerging fields need to be known and understood. Nanotechnology, unlike any other technology, can find applications in virtually all areas of human life. In spite of being an infant at its evolution, some of the known issues related to nanotechnology suggest a wide spectrum of potential societal impacts. The current public nano-discourse provides sociology with a unique opportunity to switch from

Source: www.cohesion.rice.edu. CURRENT SCIENCE, VOL. 90, NO. 5, 2006 Current Science Association. Reprinted with permission.

a merely passive, observational role to an active participating one, especially where the key players involved meet to find joint and concerted solutions for development.

The objective of this article is to address the wider societal implications of nanoscience and its deriving technologies on society within the context of the developing world. To prepare readers for the novel scientific and technological terrain, we begin with an introductory historical retrospective of the scientific and technological evolution, from the first Industrial Revolution to the third and onwards to the newly emerging technologies of the 21st century, to illustrate the influence of technological progress on the evolution of the society. The rapid pace of discoveries and development in the realm of nano indicates that this newly emerging field is different from others and therefore gives rise to specific questions, although there are problems common to other advanced technologies. The aim of exploring the world of nano is to discover new properties and to translate new knowledge into the manufacturing process for obtaining enhanced structures and components with novel chemical, physical or biological properties. This futuristic manufacturing method can virtually invade and pervade all areas of human life, since it modifies the identity of all matter, animate and inanimate. It will, for sure, revolutionize human society in an ever unprecedented manner. Certain is that this new scientific branch and associated technology will generate desired but also undesired results, which will have consequences for the society at large and change its structure, organization and functioning in the longer term. Different aspects relevant to nanoscience and nanotechnology needs to be woven into the nanodiscourse and thus, we are looking into these specific aspects with focus on the developing world. We try to provide a comprehensive framework on issues relevant to nanoscience and technology and map the social, economic, political, legal and ethical aspects, which are applicable to the developed as well as the developing world. The emphasis in nanoscience and nanotechnology, hailed as the science of the future and the technology of the next generation and believed to possess infinite market potential, lies on the control, manipulation and construction of matter at the atomic and molecular level. Recent studies have revealed the significance and importance of nanotechnology for fostering economic growth, human health and incrementing wealth in the developing world. Possible applications of nanotechnology in the fight against poverty will be addressed and illustrated with practical examples.

From the first Industrial Revolution to the Nano Revolution

A Historical Retrospective

When James Watt invented the steam engine in 1765, he could not have imagined that his invention would unleash the first Industrial Revolution and transform society in a way only the invention of the wheel did in 3500 BC. The steam engine entered the history of technological evolution as the symbol of the era of mechanization. Transformation of the transportation sector, the advent of capitalism, rapid urbanization and gradual rise of the modern industrial society are landmarks of that period. The rapid expansion of the industrial sector, successive inventions and technological breakthroughs enabled extension of the factories to technology parks leading to increased production and all of these prepared the terrain for the next revolution to come. Industrial Automation, also known as the second Industrial Revolution started around the year 1870. Industrial robotics transformed the technological manufacturing process, since it enabled automated large-scale production at reduced costs. Introduction of the assembly line during the 20th century changed the whole organization of the manufacturing process and labour. Improved and more efficient production methods made mass production at a global scale possible. To satisfy the insatiable hunger of the post-modern society and upcoming consumer society, mass production at an ever-increasing productivity rate and speed became a must. Invention of the transistor and the gradually growing semiconductor industry paved the way for the third revolution, the Digital Revolution. Computerization, starting in the early 60s led to global large-scale production, higher cost efficiency and introduced flexibility into the manufacturing process. The era of post-modernism had just begun when in 1959, Richard Feynman delivered his speech, 'There is plenty of room at the bottom'. Most likely, he had imagined that his prognostics would lead to further transformations of the manufacturing process although he might not have thought of a new revolution, the nano-revolution in the new millennium. The technologies of the late 20th century and newly emerging technologies of the 21st century, namely information and communication technologies and biotechnology, have notably altered the industrial and service sectors, the production methods and society. They inaugurated the information age and biotechnology era and have given way to a new kind of society, the knowledge society. Since nanoscience and technology as well as manufacturing of nanomaterials are no longer futuristic fantasies but reality, we should inquire into how the 'disenchantment of the atomic world' will change our present societies. What comes next, the nano-society?

Milestones on the Trajectory of Nanotechnology

The cornerstone of nanoscale science, and technology was laid when Feynman envisaged the possibility of arranging atoms to create new matter at the atomic and molecular level. In 1964, Glenn T Seaborg, Nobel Laureate in Chemistry, patented two of the elements, americium and curium. This was the beginning of patenting atomically and molecularly engineered matter. The term nanotechnology was coined in 1974 by Norio Tanigutchi, professor at Tokyo Science University, who referred to precision manufacturing at the scale of nanometres. In his MIT doctoral thesis of 1981, Eric Drexler extended the term and studied the subject in depth. During the same year, Gerhard Binnig and Heinrich Rohrer were, awarded the Nobel Prize in Physics in 1986 for inventing the scanning tunnelling microscope, a novel measurement tool allowing the sensing of matter at nanometre scale. This was a significant technological breakthrough and therefore had a great impact on the future development of nanoscale science. In the early 90s, Warren Robinett, University of North Carolina and R Stanley Williams, University of California established a virtual reality system and linked it with the aforementioned scanning tunnelling microscope to 'see' and 'touch' atoms. When D M Eigler placed xenon atoms to a shape, reflecting the logo of IBM in 1990, he rendered the proof of the possibility and feasibility of atomic-scale manipulation and this marked a further pioneering breakthrough in this newly emerging field[1]. In 1993, the first academic research centre dedicated to nanotechnology was institutionalized in USA, namely at the Rice University. Only five years later, Zyvex, the first molecular nanotechnology company, was established in USA and that marked the beginning of private nanotechnology venture capital companies. In 2000, one further step ahead was made, when Lucent and Bell Labs, together with Oxford University created the first DNA motor, the first nano-biotechnological gadget. Since the beginning of the 21st century, molecularly precise manufacturing, nano-factories, public and corporate investments in nano-research and public-private venture partnerships are rapidly expanding and increasing. The development in this newly emerging area suggests that the time-span between milestones on the trajectory of nano-technology is reducing considerably, the further we move on. Nano has gathered momentum and the nano-revolution has just begun.

The Distinctness of Nanoscience, Engineering and Technology

We have seen how the scientific discoveries and technological inventions of the past three centuries have revolutionized the manufacturing sector and society. The common traits of the new technologies that transformed human and social life between the first Industrial Revolution and the Digital Revolution were that they were designed for large-scale production, which was capital and energy-intensive, in need of large manufacturing infrastructures and competing in the global markets. Big became beautiful and the manufacturing processes and methods were aligned along the macro principle. The transistor and semiconductor industry inaugurated a change in that trend, and small became beautiful and manufacturing began to focus on micro and then on nano.

Nanoscience and technology requires nanomaterials. Manufacturing of nanomaterials is not just a step further down in size; it is about using the knowledge of the atomic realm to produce novel artefacts in a cheaper and cleaner way, with reduced capital and energy inputs and with more precision. The unparalleled development of nano-technology and the dissimilar preconditions for nano show that nanoscale science and technology is different from the precedent and other newly emerging sciences and technologies. We highlight a few of the most salient peculiarities that explain the distinctness of nanoscience and manufacturing of nanomaterials. First, at the nanometre scale, science and technology converge and therefore go beyond the traditional boundaries of disciplines. Nanoscience is of trans-disciplinary nature since it involves chemistry, physics, mathematics, cognitive science and life sciences, in particular genomics and proteomics. Nanotechnology fuses with other recent technologies like information and communication technologies and biotechnology. Second, control and manipulation of the very elementary building blocks of all objects of the living and non-living world – atoms and molecules – enable modifications of the same, which can influence every area of life. In other words, the core novelty in science and technology on the scale of the nanometre is that scientists and technologists do not invent the world *ex novo*, as in the past, but *de novo*, since the new artefacts are made of components which have no natural analogues. Third, the term nano refers to measurement, the nanometre as it indicates the size of the matter being observed and manipulated and the term does not refer to any object per se. This explains the unlimited spectrum of nano since all physical matter, irrespective of its nature, can be measured, and the only condition is that measurement facilities for that size regime exist. Fourth, manufacturing of nanomaterials does not need the enormous initial capital outlays for industrial infrastructure that other technologies require and therefore gigantic technological parks have become obsolete, at least in cases where shape control of the nano object is not a stringent condition. Factual examples substantiating the assumption that production can be cost-effective and tailored to local needs – either large or small – are available. These examples include the production of nanomaterials through biology. We will come back to this argument later. A further novel aspect with regard to nanotechnology that gives a practical expression to the pace of technological change is the reduction in time from the scientific discovery to the application of the new knowledge.

Implications of Nanoscience and Nanotechnology on Society

Science and Technology Changes Society

In order to understand the nexus between nanoscience, nanotechnology and society in the context of social change, a few preliminary concepts having general character and hence, applicable to other technologies as well, will be briefly outlined. Science, technology and society are intrinsically interlinked and characterized by mutual interdependency and since technology is firmly imbedded, it cannot be looked in isolation. Application of scientific knowledge and associated developments are two of the major factors determining social progress and prosperity. Social change is as dynamic and complex as social systems are and with regard to social evolution, it is both the essential ingredient and the driving force. A myriad of factors determine the technological, social and cultural evolutionary processes of the society. Advances in any discipline inevitably lead to changes in social relations, meanings and societal patterns. Earlier, we have seen how scientific discoveries and technological inventions had literally revolutionized societies with time. Considering the time factor, technological and social changes may not occur contemporaneously, since the social system requires its own time to respond to alteration and to find its new equilibrium. Today's coexistence of several forms of societies and the analogous coexistence of technologies, from pre-industrial to state-of-the-art technologies across the globe again illustrates the above facts. Progressive technological and social changes do not necessarily eradicate previous forms and historically, science and technology have been used by all kinds of societies irrespective of their stage of development and belief. With regard to social and technological change, as far as nanoscience and its deriving technologies are concerned, it is likely that their potential impacts will be stronger, because nano has the incomparable force to pervade all societies and economies, from the pre-industrial to knowledge societies, from ancestral to highly industrialized economies and is not necessarily subjected to a nation's current development stage and/or geographical location. Nanotechnology has the asset of both, to bypass and to bridge yesterday's missing technological link between the developed and the developing world.

Society and the Scientific and Technological Innovation Process: Protecting, improving and preserving life using new findings is intrinsic to scientific and technological research, since the very premise of science is to serve humanity. Society reacts to technological change with new forms of institutions and develops its own responses to technological innovation, and that is valid for nanoscience and technology as well. Similar to all other cybernetic systems, a change in one of the systems will generate alteration in the others because these complex organic systems are linked by multiple feedback loops and therefore the social world always responds to technological innovations. The innovation process shapes the evolution of society, and therefore it is essential to understand the societal implications of nanoscale science and technology in order to know and understand in what direction society is advancing. The current infant stage of this technology limits reliable or accurate prognostics; but in spite of these limitations, a historical retrospective on how technologies of the yester centuries revolutionized the social organization, structure and value systems can help understand and predict the potential impacts of nanoscience and nanotechnology.

Forecasting the Nanofuture: How will society respond to the distribution and diffusion of engineered nanomaterials, including commodities, instrument facilities and services and how will these change society? As nanoscience and nanotechnology move forward, it has become a necessity to ask such questions. Commonly, it is believed that the social implications in the developed as well as in the developing world will be similar to other newly emerging technologies of the 21st century like biotechnology and information and communication technologies. This assumption neglects two facts: first, nanotechnology is a fusion technology and therefore incorporates, for instance, bio and information technologies. Synergy effects will amplify the complexity and inevitably exceed the hypothetical consequences of one single technology. Second, the world is going into nano, even where information and communication technologies yet have not pervaded society at large. In developing countries, where pre-industrialized and post-modern forms of technologies coexist with newly emerging technologies, nano-engineered commodities and services can be designed for the needs of people belonging to pre-industrialized, post-modern or knowledge societies, since no preclusions apply. As far as predictions of the future of nano are concerned, global trends suggest that it is gathering momentum. Expansion in research and development, public and corporate investments, public-private partnerships, media coverage, patents, services and devices all clearly indicate that nanotechnology is on a sharp rise. From these positive projections one can conclude that nano has the potential to become the flagship of the industrial production methods of the new millennium in developed as well as in the developing world. Nanotechnology will certainly not replace all other technologies, but coexist and borrow from the technological inventions of the past. Thus, it is unlikely that the nano era will replace the digital era. It is more realistic that the digital age will converge with the nano and their synergy effects will lead to fundamental and irreversible alterations of cultures and institutions, societal organization, mechanisms and patterns, including the demographic structure. In view of its pervasiveness, it is likely that the magnitude of this new technology at the frontiers of discovery will exceed those of precedent technologies because the intensity of the impact of a phenomenon is positively correlated to its pervasiveness. These, up to now known circumstances suggest that the possible impacts of nano-technology will even go beyond those of the first Industrial Revolution.

Issues – An Outlook

Artificial Evolution – How Green is Green Nanotechnology?

The term 'green nanotechnology' apparently seems a paradox per se, as it challenges both nature and the ecosystem, because as a result of controlling and manipulating matter at its very elementary level, new matter not present in the realm of nature is created. Since the concept of 'green' refers primarily to environmental protection and not to the evolutionary process, nanoscience and technology are not inconsistent with 'green'. 'Green nanotechnology' has in fact the potential to play a pivotal role in the struggle against the world's most pressing environmental problems. Bio-nanotechnology, for instance offers a wide spectrum of new possibilities for mitigating the adverse effects of environmental degradation, having indisputably many causes and sources. In particular with regard to soil, air and water pollution and the unsustainable exploitation of natural resources, bio-nanotechnology can provide viable solutions (e.g., support of cleaner production methods, provision of alternative and renewable energy sources, reduction of input into the manufacturing process, purification of water, etc.). The interface of bio and nanotechnology however, does not only generate positive results. The potential risks inherent to the convergence of life sciences and nano-science and the deriving technologies need to be addressed and understood. Threats may arise from the increased chemical reactivity of materials at the nanoscale, the toxicity of nanoparticles and the yet unknown side effects of the atomic and molecular engineered materials. The release of atomic and molecular engineered matter into the biosphere poses additional problems to human beings and nature, since test results obtained in the laboratory may differ from those carried out in an open environment. Yet it is unknown, how humans and the environment will respond with regard to the distribution and accumulation of novel materials. The lifecycle of products containing nanoparticles is difficult to establish, since the degradation process of nanomaterials and components is only estimated. The identified hazards with regard to health are chiefly related to the absorption of nanoparticles by the human body and their distribution as well as the risk of accumulation in organs. It is further unknown, how the human (and animal) metabolism will react to the intake of nano-engineered food and nanoparticles, since once dispersed in the ecosystem, these will enter the food chain. Research into the downsides of bio-nanotechnology is essential and transparent communication of the results will, in the near future, become of crucial importance in view of the credibility and plausibility of green nanotechnology. However, recent results obtained suggest that the benefits far outweigh the risks (e.g., applied nanotechnology techniques for water purification systems).

Crossing Land – Melting of the Traditional Boundaries of Natural and Human Sciences: The realization of the possibility to explore and control the world at the nanometre scale has given life to new scientific fields, and blurred the traditional boundaries of disciplines and even led to fusion and convergence amongst natural sciences. These new circumstances present a unique chance for the sciences, hard and soft, to meet and to overcome C P Snow's paradigm of the two cultures, dividing the scientific from the human sciences[2]. Despite the scientific tradition of hard and soft sciences to use different methodologies and jargons, at the nanoscale there seems to be a trend of convergence of the two apparently opposing scientific cultures. Scientists belonging to the first category have recognized that at the atomic and sub-atomic level an organic world view becomes extremely useful since observation, control and manipulation of matter in realms inaccessible to human's ordinary senses not only demand new instrumental facilities but also novel approaches[3]. This revolutionary shift in the scientific mentality of natural scientists has the potential to overcome both the two culture paradigm postulated by Snow and the long-lasting *Methoden-streit* (disputes about methodologies) between the natural and human (cultural) sciences. This new situation frees the various disciplines from their life in isolation and gives natural and human scientists a chance to meet. That does, of course not mean that a physicist becomes an ethicist or vice-versa, but both needs to find a solution that contribute to the qualitative enhancement of human life. Recent conferences held on nanoscience and technology bear witness to the paradigm shift and revealed that the scientific fraternity has recognized the importance of bridging the cultural gap. Today's scientists seem to be prepared and willed to cross the borders of their own disciplines.

Education and Training in Nanoscale Science and Technology: Frontier science and technology represents a true opportunity to enhance a country's qualitative and quantitative level of human capital, since altered ways of manufacturing require the development of new capabilities and skills and that creates new job opportunities.

Enhanced human capital strengthens a country's competitiveness, spurs economic growth and prosperity, all essential ingredients for a more sustainable economic, human and social development. These factors are of paramount importance to developing countries, since they are rich in human capital. In recent years, academia has begun to react to the upsurge of the nano-phenomenon and started preparing the future workforce for the emerging opportunities arising in the nano realm by offering multidisciplinary curriculum that complement basic natural science education with specialized courses in nanoscience, materials science and molecular biology as well as sponsoring continuing education and training. Since the socio-economic situation of a country conditions its public expenditure in education, research and training, most educational and scientific activities in nanoscale science, engineering and technology are offered in highly industrialized countries. Since nano-manufacturing is possible in developing countries, it is vital to develop the necessary human intellectual resources for nano-manufacturing in the developing world. Educational, research and training programmes in nano-science and its related fields of application, analogous to those of highly industrialized economies, have become a necessity. That this is possible in the developing world as well, can be illustrated by the existing strategic partnerships between government agencies, industries and businesses established for starting nanotechnology research centres of excellence. It is widely shared that scientific education, training and research have the potential to narrowing the gap between the developed and the developing world, even if not within the society itself. It is a matter of fact that if the developing world does not catch up with the scientific progress of the industrialized world, the risk of being further marginalized becomes real and it would aliment the public fear of a new gap between the haves and have-nots: the nano divide.

The Nano Economy: Technology has often played a central role in wealth generation and the emerging nanotechnology market has the potential to transform and reshape all economic sectors, from the primary to the tertiary. The assumption that nanoscience and its deriving technologies will change the world must stem from a basic premise, namely that novel findings and innovations in the nano-metre scale will have a visible and significant impact on productivity. The projected consequences of rapidly advancing technologies on the nanometer scale will see the rise of new industries and the fall of those stuck to conventional or sustaining technologies of the past[4]. Marketing of scientific discoveries at the frontiers of science and technological innovations in the nano realm will become the driving force of the nanomarket. We can assume that in the years to come, this relatively new economic phenomenon will contribute to an increase of the global Gross Domestic Product (GDP) as projected by global economic institutions like the World Bank and International Monetary Fund. A comparison of the economic performance of the nano industry and business with the development of the national and global GDP could reveal possible correlations between the two and provide an answer to the question if nanotechnology truly contributes to economic growth in quantitative terms.

The Nanobusiness and Finance: Since the nanobusiness is still in its infancy even in the developed world, it is difficult to assess its possible economic impacts in the developing world. In highly industrialized economies, like USA, commercial spin-offs of nanoscientific research, start-ups and venture capital enterprises, together with multinational companies reaching out for nanotechnology and competing for market shares of the newly emerging nanotechnology market will shape the national and world economies. Still in its nascent phase but rapidly growing, the nanobusiness encompasses a wide spectrum of well-established and solid manufacturing branches, from biotechnology, materials, electronics, energy, healthcare, textiles, sensors and many others and does not preclude any business segment. The ongoing globalization process tends to amplify the importance of nanotechnology use, since manufacturing of nanomaterials provides the link to international capital markets and global technology and production networks. Not only is the future nano-market impressive in size, so also the potential clientele of nano-technology derivates, products, devices and services. Since every matter is ultimately composed of atoms and elements, theoretically all people could become exposed to the nearly unlimited nanomarket and become its benefactors. To put it into perspective, the US National Science Foundation (NSF) estimates[5] the total global market for nanotechnology-related products and services to reach US $1 trillion by 2015. Based on these projections, the Nanobusiness Alliance forecasts[6] the global nano-technology market to reach US $225 billion by 2005 and expect it to be US $700 billion by 2008. By confronting these figures with the projected world GDP, the share of the nanomarket[7] will exceed 1% by 2008. The ever-increasing number of private nanotechnology companies and the sharp rise in patent filings from public and private institutions testify the upward trend in commercially exploiting innovations made in the field of nano-manufacturing and the run into nanobusinesses. To illustrate what is stated before, the US Patent and Trademark Office issued until the first half of 2005, 3818 patents with reference to nano and 1777 patent applications were handed in and awaiting to be transformed into lucrative licences[8]. The private nanobusiness was born only in 1997, when nano pioneers started the first venture capital company in USA. Not only large and well-established trans-national companies (TNCs) followed Zyvex and started diversifying and integrating attributes of nanotechnology techniques into their manufacturing activities, but also small and medium sized enterprises (SMEs) entered the arena of the upcoming nanobusiness. Preliminary findings of business experts suggest that the latter have already conquered niche markets and by consolidating their positioning at the global level, these challenge the big market leaders. A look at the companies and sponsors present at the 8th Nanotech Conference and Trade Show in May 2005 in Anaheim, California reveals who the contenders in the newly emerging market are. More than 100 corporate companies, both TNCs and SMEs are in some way involved in nanomaterial manufacturing[9]. As far as the capital market is concerned, nano-stocks are already traded at the world's major stock exchanges and financial institutions as well as insurance companies nurture vast interest in this novel technology and its derivates. With regard to private investments, and risk management they already play a significant role. At the micro-economic level it is still difficult to express prognostics because of two major reasons, namely the absence of hard data about the cost structure of nano-engineered products, devices and services and secondly, nanotechnology commodities are yet not an integral part of everyday life in any part of the world (primarily because there are only a few commodities for mass consumption).

Public and Private Investments in Nano Research and Development: Public and private research funding in nanoscale science and technology has progressively increased over the past years in both the developed and developing world. Research and development activities are as yet generally segregated in relatively large industrial, government and academic laboratories; but the latest trend suggests, to a smaller but not less remarkable extent, that the private sector and in particular SMEs, are investing into the nanometer scale technology. The public sector still holds the lion's share of research and it has been growing at an unprecedented rate in the last years. The global public spending in nanoscience and technology exceeded US $3 billion in 2003 and it will increment, since more countries, including developing countries are planning or have already launched national nano-initiatives[5,10]. A look at the government funding for the next five years in USA confirms the progressive trend in public investments into nanoresearch and development. For the year 2005, US $809.8 million has been approved and the figure for 2008 exceeds an annual spending of US $ one billion[11]. Developing as well as newly emerging economies have started realizing the inherent opportunities of nanoscience and nanotechnology and the importance to compete from the beginning and not, as it was the case with information and communication technologies, to wait that these will be transplanted from the developed world at a later stage. For a good performance in the global R&D arena, public and private investments into nano is a precondition. The trend of increasing public spending into nano in the developing world goes parallel to that of highly industrialized; the only reservation is that their investments are lower than those of the developed world, if compared to the respective GDPs.

Reliability, Safety and Risks – Assessment and Management: Systematic identification and assessment of the risks of a new technology are essential. The idea of a system getting out of control is the nightmare of every conscientious scientist and technologist and not only of techno-phobes and those predicting a cataclysmic end of humanity, because of the release of toxic nano-engineered artefacts from the laboratory into the environment. The potential risks might arise as a result of the characteristics of the nanoparticles themselves, the properties of products manufactured with nanoparticles along with the manufacturing process. Systematic research into the downsides of this new technology, especially into the risks and perils is essential, chiefly because of two reasons. First, because if unintended consequences, including negative side effects and counter intuited results are known, it is possible to calculate the risk and to take precautions. The development of worst-case scenarios which include the necessary mechanism and measures for managing such striking situations serves to limit or mitigate the adverse effects on society and its natural habitat. Several existing instruments and methods used in technology management, like the Environmental Impact Assessment (EIA) help in evaluating the reliability, safety and risks of new products, services and devices using atomic or molecular engineered matter. Secondly, the fear of derailing nanoscale science and technology, because it is evolving faster than the researchers' ability to keep pace with the development, independent of whether it applies or not, is undermining the research and development process and gives rise to anti-nanotechnology feelings and attitudes. Therefore, it is indispensable to foster methodical research in the fields of safety, reliability and risk management of nanocommodities, devices and services. The apocalyptical worst-case scenario involving nanotechnology, named by Eric Drexler, as the 'Grey goo', has contributed little towards broadening our societal understanding and knowledge of the potential risks of molecular engineered artefacts[12]. Since people are more receptive to fictional representations as they shape their imaginaries, the 'nanomania' has rather alimented the entertainment and leisure industry, giving rise to misconceptions, misunderstandings and distorted views regarding this particular technology rather than contributing to a rational and genuine discussion.

Nano Policies and Institutions: When we look at the nano-phenomenon from the political perspective, in general we look principally at issues relating to long-term strategic policies, including intellectual property reforms, international cooperation, monitoring and regulation of research and development. Because of its unlimited potential, nanoscale science and technology requires the adoption of national and international policies. The management of research and development in nanoscience and technology, in terms of administration and control by the public authorities is to guarantee for the responsible use of the potentials both exemplify. The claim for formalizing a regulatory regime over nanoscience and technology reflects that a minimum of coercive intervention by the state is needed, especially vis-à-vis responsible nanotechnology. Scientists, politicians, the private sector and representatives of the civil society have opposing views as far as the intensity of the public interference into scientific and technological research and development is concerned and even within these groups the views do not converge. The opponents, the nano sceptics, an utmost heterogeneous flock, call for a regulatory and institutional framework that limits the scope of scientific activities and the diffusion of atomically and molecularly engineered commodities and services. Despite the competing interests and colliding opinions, the two blocks agree on three central issues in the public nano discourse, namely the non-interference into privacy, preservation of human dignity and protection of the society and natural environment from hazards. The first two, since they have strong ethical connotations, will be addressed apart and the third has already been addressed. To counter misuse, including activities and actions that are in net contradiction with the universally shared ethics and principles, a minimum of a regulatory framework has become a necessity to guarantee responsible use of nanoscale science and its deriving technologies. The current scenario suggests that the institutionalization of a supranational body with a standing committee exercising a minimum of international control, monitoring the technological development in the atomic realm and providing a legal framework to which the countries doing nano research and development conform their activities, makes sense. Despite differing value systems, political and legal traditions and systems, the international community should agree at least on a minimum political control and administration in order to ensure that nanoscale research and development is consistent with the ethical principles present in all universal value systems. Defining the perceived benefits and risks of nanoscale science applications, as perceived by the scientific community, industry and business and public, becomes valuable while policymakers decide what is best for their community and the nation.

Nano rules and regulations: The laissez-faire approach of the past has led to a lack of rules and regulations regarding research, development and deployment of atomically and molecularly engineered products. The absence of appropriate norms at both, the international and national level, reflects what W F Ogburn described as the cultural lag[13]. According to his theory of technological evolutionism, there is a gap between the technical development of a society and its moral and legal institutions. It is a social fact that while scientific and technological development in nanomaterials manufacturing is leaping ahead, legal regulations lag behind and open a gap of trust between the people and public authorities. The present situation has the potential to generate social tensions and problems since misuse of nanotechnology for destructive purposes is yet not ruled out explicitly. There is a need for new forms of regulations and international standards to direct research, development, manufacturing and commercialization of nano-technology. However, the mere establishment of a legal framework or standards is not enough, since the laws and regulations need to be enforceable.

A better understanding of the legal implications of nanotechnology would help policymakers in the establishment of a regulatory framework and standards that are consistent with the dominant value system of their society. This is particularly valid and important for developing countries, where during the colonial and post-independence times new laws and institutions had been introduced, without taking into consideration the local culture, institutions and traditions. Since control and manipulation of the genetic signature imprinted in the DNA of each human being has become possible and manipulation of human cells, including stem cells is done in laboratories, a regulatory framework setting limits to the exploration of human nature is a necessity. However, none of the regulations should delay or inhibit the growth of knowledge aimed toward the betterment of humankind.

Nano ethics – A deontological code for the nano-community: Not all what can be done, should be done; this is the nucleus of the ethical dimension with regard to research and the technological innovation process in general, and it applies to nano in particular. Since science and technology on the scale of the nanometre is not restricted to the domain of materials science but reaches out to life sciences, it is important to understand the ethical implications. The formalization of a deontological code is needed to prevent nanoscience and technology from derailing and heading into the wrong direction. Ethical guidelines related to research procedures and activities in the realm of atoms and molecules are in fact needed to bridge the present gap between science and ethics. One of the core tenets of such a deontological code is the observation, respect and protection of human dignity and noninterference into individuals' privacy. From a purely scientific point of view, humans do not enjoy a higher status than other living creatures but from the ethical perspective, humans have a different status because of their highly complex nature due to the genetic endowment and consequently possess faculties other living beings do not have. This is where ethics and science clash. Little systematic research into the ethical consequences of nanotechnology has been undertaken so far. The ethical community needs to be actively involved into the nano debate because the expertise of ethicists is required for softening the frontiers of the two blocks. The big ethical controversies will not focus on the threats to the dignity of the average grownup adult, but will be concerned with foetuses, children, terminally sick, elderly and disabled, all belonging to the most vulnerable groups of society.

Nanotechnology and War – Nano Arms Race

The military plays a crucial role in the scientific and technological innovation process, and this applies to discoveries and development in the nanometre scale as well. In the developed as well as the developing world, large investments are made into national security and defence. Especially in the current world-political scenario, the strive for more powerful instruments to secure the national integrity and interests is a common phenomenon, independent of a country's economic status. The use of nano-engineered materials for weaponry and the potential use of atomically manufactured matter and devices for weapons of mass destruction is not science fiction, but reality. In view of today's increased political instability and diffused feeling of uncertainty, the safeguard of a nation's interests has gained importance. Therefore, it is possible that weapons equipped with nanotechnology will be deployed in present and future armed conflicts, since there is yet no international or national ban on weapons using nanoparti-cles and nanotechniques. It is the direct link between nanotechnology and ballistic studies that frightens people and also the potential abuse of nano-engineered weaponry[12]. However, we must state here that research for military use produces innovations that later could be utilized for civilian purposes as well. It is yet too early to speculate about a possible technological fallout of nano commodities designed for defence, but one cannot preclude it.

Public Perception and Public Involvement in the Nano Discourse

One of the most pressing issues in science is the involvement of the public. The triple helix, i.e. academia, government and industry, needs to be extended to a fourth dimension, the civil society, because of the latter's relevance. The public plus the print and electronic media need to be involved from the beginning, since news and information regarding nanoscale science and technology will shape the public perception and determine to a large extent people's knowledge, attitude and behaviour towards this new technology. Consumer acceptance is the key when it comes to commercially-developed nanotech-nology products because ultimately it is the end-users who will influence the trajectory of nanotechnology.

Therefore, it is of paramount importance to include the public from the very beginning, so that people understand the novelty of atomically and molecularly engineered products and services. The underestimation of consumer acceptance and consumer resentments with regard to biotechno-logically engineered organisms in the past revealed the importance of social acceptance, since knowledge and perception finally direct the people's attitudes vis-à-vis new technologies. Studies conducted in USA have already shown that artificially created matter having the attributes of living creatures, namely the ability to adapt, cooperate, learn and adjust to change occurring in their, or another system frightens people[12]. It is therefore important that such studies are conducted in the developing world as well, and especially where nanoengineered products intend to be applied. People's apocalyptical predictions with regard to nanotechnology do not find any foundation in rational reasoning. Uncertainty and unpredictability are however, effective instruments for manipulating the public opinion and to undermine trust in science and technology.

Nanotechnology and the Media: In the age of emerging knowledge society in many parts of the world, including the developing world, mass media plays a crucial role and hence shall not be neglected. With regard to nano, it is even more important to envisage the social function of mass communication tools. The print and electronic media, including the Internet, known respectively, as the fourth and fifth power, are the most influential to shape the public perception and people's understanding of nano. Scientific and technological breakthroughs find their way to the public in general through the media, and thus it is important to inform people in a way they understand. To assess the trend of the nano presence in the media, there are two possible ways of action. If we use the number of scientific publications about nano as a variable indicator over a certain time-span, we can assess the trend and trace its development over the past years and forecast future trends. The same can be done with publications in popular science, business media and entertainment. The rapid increase in publications in scientific literature over the past ten years is striking (Figure 1). To put it in perspective, in 1995, 4372 nano* references were recorded in English scientific publications; in 2000, there were already 11,447 and in the first five months of 2005 alone, there were 16,518 titles. These figures, made available by the Institute of Scientific Information (ISI) Webofknowledge, do not leave any doubt about the rise in the number of scientific publications and this tendency will certainly grow progressively. The analysis here refers only to English language publications traced by ISI and does not therefore reflect the global scenario, and further includes titles that may not be purely nano-specific, but include hybrids.

Figure 1: Increase in Nano-Related Publications in English Journals

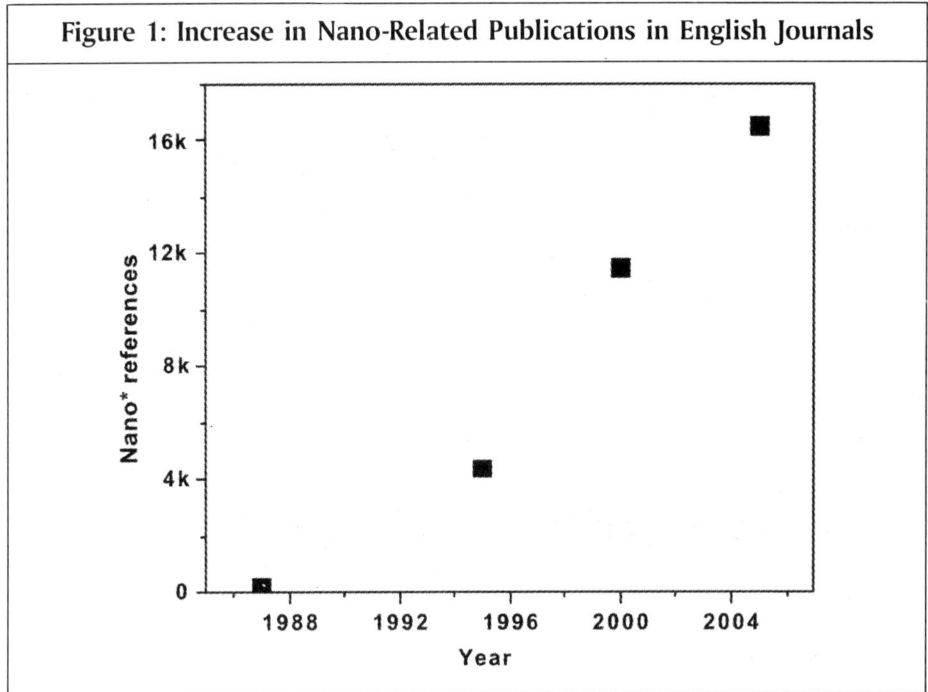

With regard to the popular science literature, business media and entertainment a similar trend stands out, although it could not be quantified. Broadcasting, television and film industries and also the computer game industry have already developed great interest in nano for increasing their clientele and turnovers. Science fiction literature about nano is not only fascinating adults, but also youth; video games have already paved the way for nano to enter children's rooms. The media hype about new discoveries in the atomic and sub-atomic world gives rise to distortions and reflects commercial exploitation of an event, rather than the aim to popularize the potentials of nanotechnology or to inform the public of the latest discoveries or achievements made. Without any doubt, the media has become a platform where scientists, government officials, pressure and advocacy groups and social activists voice their stance and forge the minds of the uninitiated audience.

The Public eye on Nanotechnology: The exclusion of the civilian society in a dialogue on the potential positive and negative implications of nanotechnology and the course on which academia, public and private research funding agencies, and the industry is navigating, will have disturbing consequences and cause a public opinion backlash. Therefore, the triple helix must become aware about the role of the public in advances in science and technology. Participation of the civil society in the current public debate on nano is without any doubt a sensitive issue. Public fears, possible negative social response or even rejection of nanotechnology indicate that the involvement of the civil society, i.e. social movements, non-governmental organizations (NGOs) and community-based organizations (CBOs) is no longer an option, but a must. Resentment and social discontent about nanotechnology have been openly voiced by the call for a moratorium on the deployment of nanomaterials and the leverage of nano-critics and nano-sceptics can no longer be ignored. The mobilization against nanotechnology of certain social activist groups is a first, but alarming sign that public acceptance cannot be taken for granted. The primary intention of the anti-nano community is to heighten public awareness on possible negative impacts of nano and influence the policymakers to jeopardize technological advancement. With regard to the developing world, it is less likely that these groups will be influential since the democratic instruments to challenge national policies are limited. Further, social acceptability is more likely, where people are willing to accept the risk of new technologies and generally have a positive view about science.

Harnessing Nanotechnology for Economic and Social Development

Nanotechnology and the Developing World

The way industrialization advanced in the developing world has no parallels in the highly industrialized countries. In the former, various kinds of technologies – from indigenous to state-of-the-art technologies – coexist and analogous to that, a plurality of different kinds of societies live together, ranging from pre-industrial to emerging knowledge societies. The ongoing industrialization and modernization trend in the developing world has generated a variety of problems that culminated in the global phenomena of environmental pollution, widespread diseases and urbanization[14]. The situation in the developing world has not significantly improved and in certain countries, the state of the people's conditions has even deteriorated. The world's most pressing problems are manifold and they project a myriad of issues. Extreme poverty, lack of education, high rates of mortality and morbidity, widespread epidemics and environmental problems predominate. Innovative and holistic approaches and strategies need to be developed and implemented when addressing these highly complex and intertwined problems, and no technology should be considered irrelevant. We stated earlier that nanotechnology can find applications in societies and economies, irrespective of their development or status. This tells us that at least theoretically, everyone could benefit from its potential applications. Nanoscale techniques have the potential to be in harmony with traditions and in synchrony with technology and this is what makes them become a valid tool in the struggle against poverty. They could indeed make a significant difference in the current scenario and contribute to a more sustainable economic and social development as we will see hereafter. Several developing countries have recognized nanotechnology as a catalyst for economic, human, social, technological and environmental development and launched national nanotechnology initiatives. Worldwide, more than one third of all nations are promoting research and development, including education and training of nanoscientists and nanotechnologists and more than seven countries belong to the developing world[15]. To put it into prospective, in India for instance, the Department of Science and Technology has allocated in the present five-year plan US $20 million for the national Nanomaterials Science and Technology Initiative. Several academic institutions in India already possess the necessary facilities to compete at a global level and become research centres of excellence, with highly educated and trained workforce, state-of-the-art research infrastructure and possible links to the industry and business. In Figure 2, we present the investment in nano research in a number of countries, grouped separately. It is clear that the investment in India is low in comparison to other countries in the region.

Developing countries are rich in human capital and their brainpower in the medium and long-term, will reshape the imbalance between the North and South. Nanotechno-logy offers a new opportunity to the manufacturing industry in the developing world. The wide range of possible applications of nanoscale technologies suggests that if the industrial sector of developing countries is involved in the manufacturing of nanomaterials, it will enhance its competitiveness in manufacturing at the global level.

Figure 2: Investment in Nano Research for the year 2003 in Some Countries, Given as a Fraction of their GDP

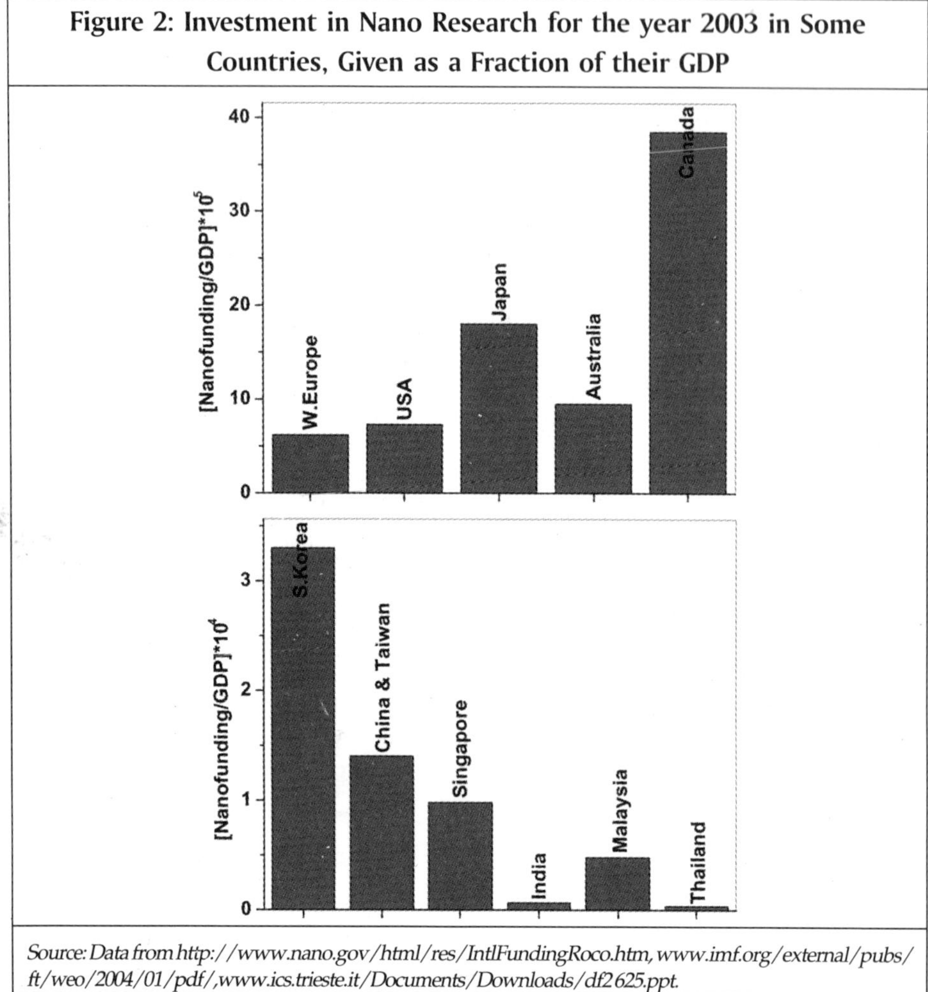

Source: Data from http://www.nano.gov/html/res/IntlFundingRoco.htm, www.imf.org/external/pubs/ft/weo/2004/01/pdf/,www.ics.trieste.it/Documents/Downloads/df2625.ppt.

Harnessing Nanotechnology for Sustainable Development: Development and application of technology, if expected to be successful, need to be designed to the needs of the target group and be suited for the socio-economic context. Nanoscale science and its deriving technologies can virtually enhance the lives of nearly everyone, the rich and the poor, because of its pervasive benefits and its suitability in resource-limited settings[16]. It has the potential to provide most innovative tools and strategies in fighting poverty-related issues. Five billion people live in the developing world and their living conditions could be enhanced by the diffusion of applications of discoveries made in this area. To substantiate our optimistic views with concrete examples, we have identified seven core areas where nanotechnology can make a significant difference in the developing world.

(a) Economic development: Nanobiotechnology, involving the biological production and utilization of nanomate-rials, is a promising new field, especially in the developing world with its unparalleled biodiversity. This asset of the developing world can be harnessed through nanomaterials synthesis using microorganisms, including bacteria, viruses, fungi as well as plant and animal-based products. Several examples of nanomaterials synthesis using biology have been reported[17]. Techniques related to nanobiotechnology do not require large investments and infrastructure and can therefore be developed on the site of application itself. The green and cost-effective solution to nanomaterials manufacturing is one example that proves that with the necessary knowledge and skills, nanotechnology can be developed and diffused in the developing world. This approach has been applied to the synthesis of nanoparticles and nanotriangles of gold as well as various other inorganic nanoparticles such as those of CdS and $CaCO_{3.}$ These methodologies can be suitably adapted for the large-scale synthesis of materials for applications such as cancer therapy, IR absorbing coatings, etc. Synthesis of nanoparticles in human cells added a new dimension to this research[18]. It is also likely that an understanding of the underlying processes may permit us to alter the chemistry such that shape and size control of nanomaterials becomes possible. The biochemical events may be transplanted to other organisms so that processes similar to the bulk production of enzymes become feasible. All of these could happen in the very foreseeable future, with lower investment than that necessary for chemical or physical routes.

(b) Safe drinking water: Among the numerous applications of nanotechnology one can visualize, the most widespread impact as far as the developing world is concerned is perhaps in the area of water purification. Access to safe drinking water is one of the major concerns in the developing world, since almost half of the world population has no access to safe drinking water and basic sanitation. Water purification systems, equipped with nanomaterials and using new kinds of membrane technologies with variable pore sizes as filters could provide people in any area with safe drinking water. These are easy in application and maintenance and already available in the market; the forward-osmosis membrane technology

of Hydration Technologies[19] is one technique utilizing nanotechnology. Thus a combination of nanotechnologies will be useful in providing cost-effective and safe drinking water, which will have less dependence on energy. Although the product is right now marketed for emergency water supply, large-scale water purification is indeed feasible. To substantiate the validity of these suggestions, we mention the following. Carbon nanotube-based filters could be developed for water purification. The development of a filter which can separate petroleum hydrocarbons from crude oil has been demonstrated[20]. The filters also remove bacteria from water. With nanotubes, smart sensors could be incorporated into the filter as several nanotube-based sensors are known already. Nanoparticles have been shown to degrade pesticides and pollutants[21]. Several nanomaterials are known to be antibacterial and they can be incorporated on various kinds of substrates[22]. It may be mentioned that we have not listed numerous other discoveries in the area related to this application.

(c) Improving food security: Nutrient deficiency is a widespread phenomenon throughout the developing world, which challenges the physical and mental health of over one billion people. Food starvation can, but not always is related to crop failure. Novel techniques using nano-technology can be applied in agriculture for breeding crops with higher levels of micronutrients, enhance pest detection and control and improve food processing. The lack of adequate storage facilities besides crop failure is one of the major reasons causing food shortage in the developing world, and in particular in remote areas. In India alone, huge quantity of wheat and rice are rotting in the open. Especially in the tropical belt, food gets spoiled easily because increased temperatures favour the growth of microorganisms, which reduce its quality or render it even inedible. Oxygen accelerates the degeneration process because it enables growth of microorganisms. It is known that carbon dioxide inhibits the growth of microbes. Carbon nanotubes could be used in food processing and preservation as an oxygen scavenger and can prevent packed food from deteriorating[23]. Another application of one of the recent nanoscale innovations involves atomically modified food and this marks the beginning of a radically new paradigm for food production. It has the potential to sidestep the controversial genetically modified food and increase the yield of agricultural produce. In Thailand, researchers at Chiang Mai University have modified local rice varieties to develop a variant that grows throughout the year

by applying nanotechnology[24]. This particular nanotechnology technique involves the perforation of the wall and membrane of a rice cell through a particle beam for introducing one nitrogen atom into the cell, which triggers the rearrangement of DNA of rice. Novel techniques applying nanotechnology could contribute to improving the current situation, which in several parts of the globe is alarming.

(d) Health diagnosis, monitoring and screening: Nanoscale techniques have the potential to revolutionize the health sector, in particular in the fields of diagnosis, screening and monitoring of diseases and health conditions[25]. A large spectrum of novel applications using nanoscale techniques in health care is possible as the beginning of a new paradigm for health care. Lack of accurate, affordable and accessible diagnostic tests impedes global health efforts, especially in remote and in accessible regions and poor settings. Widespread communicable diseases like HIV/AIDS, malaria, tuberculosis, etc., could be diagnosed with screening devices using nanotechnology. Standard diagnostic tests for widespread diseases in the developing world are costly, complex and poorly suited to resource-limited settings. A radically new approach to health diagnosis has been developed in India by the Central Scientific Instruments Organization (CSIO). Theoretical simulation and design parameters for a micro-diagnostic kit using nano-sized biosensors were completed in 2004 and it is ready for clinical trials[26]. The techniques are based on highly selective and specific biosensors and receptors like antibodies, antigens and DNA, which enable early and precise diagnosis of various diseases. The diagnostic kit 'Bio-MEMS' (micro-electro-mechanical-system) has the size of about 1 cm^2, costs around Rs.30 per piece and is easy to apply. Testing time is rapid and only requires a tiny amount of blood. This novel diagnostic tool could also find application in the detection of other diseases and pollutants in the environment, including water and food[27]. To develop therapeutics to combat malaria caused by the parasite *Plasmodium falciparum*, a common disease in many parts of the developing world, Subra Suresh and his team at MIT are using nanotechnology to systematically measure mechanical properties of biological systems in response to the onset and progression of the disease[28]. Innovative drug delivery systems using nanotechnology are another area where nanotech-nology can find application. Cancer in the developing world as widely diffused as elsewhere, is a big challenge to human health. Latest results obtained in cancer detection and treatment with nanoscale techniques provide hope that

nanotechnology could be heading for a breakthrough in defeating this disease. A way to detect cancer safely and economically is by the injection of 'molecular beacons' into the body. Britton Chance and his colleagues at the University of Pennsylvania developed tiny capsules that use the specific biochemical activity associated to a tumour to detect breast cancer[29]. As far as cancer treatment is concerned, Jennifer West and her team at Rice University, Houston, developed gold 'nano bullets' that can destroy inoperable human cancers. The nanoshells consist of tiny silica particles plated with gold and when these are heated with infrared light, the cancer cells die[30]. Carbon nanotubes have been transported into the cell nucleus and continuous near infra-red radiation absorption of nanotubes causes cell death. This methodology has been used for cancer cell destruction[31].

(e) Environmental Pollution: Environmental degradation due to unsustainable ways of production and other human activities exposes the entire world population to increased risks. Innovative techniques using nanoengineered materials and devices can be deployed for the removal of polluting molecules in air, water and soil. Cleaner manufacturing processes and methods, applying nanoscale techniques, could also contribute to lesser environmental pollution, especially in the developing world where international standards are often not safeguarded. Arsenic in soil and water for instance, is a widely diffused problem in several regions of the developing world. A simple, cheap but effective nanoscale technique to remove arsenic involves Tio_2 nanoparticles[32]. Nanomaterials have been shown to be effective in removing metal ion contamination. A wider application of such technologies harnessing discoveries in nanoscience could have a positive and wider impact on the health conditions and natural habitat of millions of people.

(f) Energy Storage, Production and Conversion: The chronic power shortage and increased need for energy resources of the rapidly growing population and economies of the developing world are challenging the energy market. Since almost all sources of energy are not renewable, soon the world will face a global energy supply problem. Solar energy is an interesting and valid alternative, especially in the sun-rich South. Scientific studies have demonstrated that nanoscale techniques involving nanotubes and nanoparticles lead to increased conversion efficiencies. Semiconducting particles of titanium dioxide coated with light-absorbing dyes bathed in an electrolyte and embedded in plastic films are cheap and easy to

manufacture and offer an alternative to conventional energy production and storage. Because of their low cost-structure, photovoltaics using nanotechnology are a valid alternative to overcome the problem of power shortage, especially in the developing world. Researchers at Nanosolar, a venture capital start-up based in Palo Alto, California are developing cheaper methods for producing photo-voltaic solar cells using nanotechnology[33]. The idea is to boost the power output of nano solar cells and make them easier to deploy by spraying them directly on the surfaces. The approach is simple and could be easily replicated in the developing world. These highly efficient solar cells can be made with a mix of alcohol surfactants and titanium compounds sprayed on a metal foil. Within 30 s, a block of titanium oxide perforated with holes of nanometre size rises from the foil. Solar cells are formed when the holes are filled with conductive polymer and electrodes are added and then covered with a transparent plastic. These are concrete examples that nanotechnology is suited for energy storage, production and conversion in the developing world.

(g) Global Partnerships: The inclusion of the South in the nano-dialogue creates new platforms and alliances between the North and South and strengthens their ties. Allocation of some of the large public scientific fundings of nanoscience and technology could be directed to developing countries in order to foster development, diffusion and dissemination of nanoscience, engineering and technology in the developing world. Global research networks of excellence enhance the value of the international scientific community. Encouraging international partnerships between the North and South, similar to the first North-South expert group meeting of nanoscientists and nanotechnologists in Trieste, Italy in February 2005 are certainly important, but also scientific exchange and alliances between countries of the developing world are becoming a necessity in view of the ongoing regionalization trends in politics and economics[34]. Global research networks, including scientific cooperation and collaboration are needed to find joint solutions for the most pressing problems of the world community but partnerships at the regional level will, in the long term, gain importance and therefore South–South nano-networks need to be envisaged.

Unexplored Biodiversity of the South – Opportunities for Bio-Nanotechnology

The development of nanocomponents that imitate or emulate natural processes can find many applications in a myriad of sectors. Protection and preservation of species in the tropical belt have become a new dimension because of bio-nanotechnology and its potential applications. The scientific exploration of the nano-bio interface becomes interesting in view of developing synthetic life-forms, manufactured organs and bio-nanodevices. Availability of the necessary research infrastructure provides the developing countries with a strategic advantage *vis-à-vis* the industrialized world, since the biodiversity is much larger in the tropical and subtropical belt as in other geographical and climatic zones. It lies in the hands of the developing countries to make use of this distinctive asset and to discover the secrets of its yet unexplored biodiversity and the particular physical properties of its living organisms. The Lotus effect, discovered by Wilhelm Barthlott and his student Christoph Neinhuis at the University of Bonn, illustrates how knowledge of what happens at the bio-nano interface can be made fruitful. Lotusan, a dirtrepellent paint is the commercialized form of the scientific discovery of the Lotus effect. Discovering the richness of the flora and fauna present in the developing world by the developing world itself with regard to bio-nanotechnology is of crucial importance. The rich cultural heritage, including the millennia old traditional knowledge in homeopathic, ayurvedic and herbal medicine together with the large biodiversity provide developing countries with a strategic advantage which the North lacks. It lies in the hands of these countries, especially those that have already adopted nanotechnology initiatives within the bounds of their national technology development policies, to make use of the unique assets for a better positioning in the emerging global nano world.

Conclusion

The potential societal implications of the scientific and technological innovation process in the realm of nano and the diffusion of the future applications of the discoveries at the frontiers of science are only partially understood and therefore need to be further explored. The uniqueness of nanoscience and nanotechnology, especially with regard to its pervasiveness into virtually all spheres of human life, explains why the magnitude of the potential impacts will exceed that of all other conventional yester-year technologies. The convergence of the newly emerging technologies of the 21st century has the potential to revolutionize social and economic development and may offer innovative and viable solutions for the most pressing problems of the world community and its habitat. However, a better understanding of the potential benefits and hazards of nanoscale science and technology is essential because it will provide policymakers with better tools to take responsible choices. Nanoscience and its deriving technologies have the potential to improve the state of the developing world, if the applications are designed and tailored to best fit the needs of the people. Nanotechnology, unlike other technologies, offers a unique chance to bridge or bypass the technological gap between the industrialized and developing world; however, a favourable terrain for their growth needs to be prepared. Therefore, joint and concerted efforts by all concerned are needed to rule out future factions and new divides.

(Birgit R Bürgi and T Pradeep are in the DST Unit on Nanoscience, Department of Chemistry and Sophisticated Analytical Instrument Facility, Indian Institute of Technology Madras, Chennai 600 036, India. The authors can be reached at birgit.buergi@env.ethz.ch, pradeep@iitm.ac.in respectively).

Endnotes

1 Eigler, D. M. and Schweizer, E. K., "Positioning single atoms with a scanning tunnelling microscope". *Nature*, 1990, 344, 524–526.

2 Snow, C. P., *Two Cultures and a Second Look*, New American Library, New York, 1963.

3 The organic *Weltanschauung* regards all phenomena in the universe as integral parts of an inseparable harmonious entity when exploring the very nature of things and is well established in social sciences and humanities.

4 Christensen, C. M., *The Innovator's Dilemma: When New Technologies Cause Great Firms to Fail*, Harperbusiness, New York, 2000.

5 Roco, M. C., "Government Nanotechnology Funding: An International Outlook", 2003, retrieved from *http://www.nano.gov/html/ res/IntlFundingRoco.htm*

6 Nanobusiness. Retrieved from *http://www.nanobusiness.org/*

7 The projected World GDP is based on the average annual growth of 4.2% and 4.1%, respectively for the period from 2006 to 2009 and 2010 to 2015. Data are taken from the World Economic Outlook 2004, published by the International Monetary Found (IMF). Retrieved from *www.imf.org/external/pubs/ft/weo/2004/01/pdf*

8 Herring, R., Nanotech patents proliferate, study quantifies huge, complicated body of intellectual property generated by entrepreneurs of tiny, complicated technology. 2005, retrieved from *http://www.redherring.com/Article.aspx?a=11866&hed=Nanotech+Patents+Proliferate§or=Industries&subsector=Biosciences*

9 "Nano Science and Technology Institute" (NSTI), *Nanotech Ventures 2005*, retrieved from *http://www.nsti.org/Nanotech Ventures2005/*

10 Hullmann, A., Nanotechnology. Europe and the world: The international dialogue in nanotechnology, 2005, retrieved from *www.ics.trieste.it/Documents/Downloads/df2625.ppt*

11 National Nanotechnology Initiative (NNI), 21st Century Nano-technology Research and Development Act, retrieved from *http://www.nano.gov/html/about/funding.html*

12 Cobb, M. D. and Macoubrie, J., Public perception about Nano-technology: risks, benefits and trust. *J. Nanopart. Res.*, 2004, 6, 395–405.

13 Ogburn, W. F., *On Culture and Social Change*, The University of Chicago Press, Chicago, 1964.

14 Hinrichsen, D. *et al.*, Population reports, meeting the urban challenge, Population Information Program, Center for Communication Programs, The Johns Hopkins Bloomberg School of Public Health, Baltimore, Maryland, Vol. XXX, Nr. 4, 2002, retrieved from *http://www.infoforhealth.org/pr/m16/m16.pdf#search='popu-lation%20city%20developing%20countries%20half%20of'*. This publication gives a comprehensive overview on the urbanization phenomenon in the developing world.

15 Brazil, China, India, Singapore, South Korea, Taiwan, Thailand, the Philippines and others have become actively involved in nanotechnology.

16 Drexler, K. E., *Engines of Creation*, Garden City, Anchor Press/ Doubleday, New York, 1986.

17 See, for example, Southam, G. and Beveridge, T. J., The occurrence of sulphur and phosphorus within bacterially derived crystalline and pseudocrystalline octahedral gold formed *in vitro*. *Geochim. Cosmochim. Acta*, 1996, 60, 4369–4376; Klaus, T., Jo-erger, R., Olsson, E. and Granqvist, C. G., Silver-based crystalline nanoparticles, microbially fabricated. *Proc. Natl. Acad. Sci.*, 1999, 96, 13611–13614; Mukherjee, P. *et al.* Bioreduction of $AuCl_4^-$ ions by the fungus, *Vitricillium* sp. and surface trapping of the gold nanoparticles formed. *Angew. Chem. Int. Ed. Engl.*, 2001, 40, 3585–3588; Nair, B. and Pradeep, T., Coalescence of nanoclusters and formation of sub-micron crystallites assisted by Lactobacillus strains. *Crystal Growth Design*, 2002, 2, 293-298.

18 Anshup, J. *et al.*, Growth of gold nanoparticles in human cells. *Langmuir*, 2005, 21, 11562–11567.

19 See *http://www.hydrationtech.com*

20 Srivastava, A., Srivastava, O. N., Talapatra, S., Vajtai, R. and Ajayan, P. M., Carbon nanotube filters. *Nature Mater.*, 2004, 3, 610–614.

21. Sreekumaran Nair, A. and Pradeep, T., Halocarbon mineralization and catalytic destruction by metal nanoparticles. *Curr. Sci.*, 2003, 84, 1560–1564 and Sreekumaran Nair, A. and Pradeep, T., Indian patent application.

22 Prashant, J. and Pradeep, T., Potential of silver nanoparticle-coated polyurethane foam as an antibacterial water filter. *Biotech-nol. Bioeng.*, 2005, 90, 59–63.

23 Hechman, M., Better eating through nanotech. 2005, retrieved from *http://www.extremenano.com*

24 ETC Group Jazzing up Jasimine: atomically modified rice in Asia? 2004, retrieved from *http://www.etcgroup.org/article.asp?* newsid=444.

25 *Nanomedicine – Global technology developments and growth opportunities*, Frost & Sullivan, Farmington, USA, 2004.

26 *AZoNanotechnology*, tuberculosis diagnosis kit based on nanotechnology, 2004, retrieved from *http://www.azonano. com/details.asp?ArticleID=368*

27 The same organization is developing a diagnostic kit for tuberculosis.

28 Suresh, S., Spatz, J. P., Micoulet, A., Dao, M., Lim, C. T., Beil, M. and Seufferlein, T., Single-cell biomechanics and human disease states: gastrointestinal cancer and malaria. *Acta Biomater.*, 2005, 1, 15–30.

29 Schewe, P. F., Stein, B. and Riordon, J., Molecular beacons for cancer. The American Institute of Physics, *Bulletin of Physics News*, Number 531, 22 March 2001.

30 Hirsch, L. R. *et al.*, Nanoshell-mediated near-infrared thermal therapy of tumours under magnetic resonance guidance. *Proc. Natl. Acad. Sci. USA*, 2003, 100, 13549–13554.

31 Kam, N. W. S., O'Connell, M., WiWestm, J., Lsdom, J. A. and Dai, H. J., Carbon nanotubes as multifunctional biological transporters and near-infrared agents for selective cancer cell destruction. *Proc. Natl. Acad. Sci. USA*, 2005, 102, 11600–11605.

32 See for example, Pena, M. E., Korfiatis, G. P., Patel, M., Lippincott, L. and Meng, X. G., Adsorption of As(V) and As(III) by nanocrystalline titanium dioxide. *Water Res.*, 2005, 39, 2327–2337.

33 Future Pundit, "Nanotech start-ups pursuing cheaper photovoltaic solar power". 2004, retrieved from *http://www.futurepundit.com* First published in *MIT's Technology Review*.

34 The first North–South dialogue on nano was jointly organized by the International Centre for Science and High Technology and the United Nations Industrial Development Organization (UNIOD). For details visit *http://www.ics.trieste.it/Nanotechnology*

13

Nanotechnology and the Developing World

Fabio Salamanca-Buentello, Deepa L Persad, Erin B Court, Douglas K Martin, Abdallah S Daar and Peter A Singer

Many developing countries have launched nanotechnology initiatives in order to strengthen their capacity and sustain economic growth. However, to our knowledge, there has been no systematic prioritization of applications of nanotechnology targeted towards these challenges faced by the 5 billion people living in the developing world. This article attempts to convey three key messages. First, it shows that developing countries are already harnessing nanotechnology to address some of their most pressing needs. Second, it identifies and ranks the ten applications of nanotechnology most likely to benefit developing countries, and demonstrate that these applications can contribute to the attainment of the United Nations Millennium Development Goals (MDGs). Third, it outlines a way for the international community to accelerate the use of these top nanotechnologies by less industrialized countries to meet critical sustainable development challenges.

Nanotechnology can be harnessed to address some of the world's most critical development problems. However, to our knowledge, there has been no systematic prioritization of applications of nanotechnology targeted toward these challenges faced by the 5 billion people living in the developing world.

In this article, we aim to convey three key messages. First, we show that developing countries are already harnessing nanotechnology to address some of their most pressing needs. Second, we identify and rank the ten applications of nanotechnology most likely to benefit developing countries, and demonstrate that these applications can contribute to the attainment of the United Nations Millennium Development Goals (MDGs). Third, we propose a way for the international community to accelerate the use of these top nanotechnologies by less industrialized countries to meet critical sustainable development challenges.

Developing Countries Innovate in Nanotechnology

Several developing countries have launched nanotechnology initiatives in order to strengthen their capacity and sustain economic growth[1]. India's Department of Science and Technology will invest $20 million over the next five years (2004-2009) for their Nanomaterials Science and Technology Initiative[2]. Panacea Biotec (*http://www.panacea-biotec.com/products/products.htm*) (New Delhi, India) is conducting novel drug delivery research using mucoadhesive nanoparticles, and Dabur Research Foundation (Ghaziabad, India) is participating in Phase-1 clinical trials of nanoparticle delivery of the anti-cancer drug paclitaxel[3]. The number of nanotechnology patent applications from China ranks third in the world behind the United States and Japan[4]. In Brazil, the projected budget for nanoscience during the 2004-2007 period is about $25 million, and three institutes, four networks, and approximately 300 scientists are working in nanotechnology[5]. The South African Nanotechnology Initiative (*http://www.sani.org.za*) is a national network of academic researchers involved in areas such as nanophase catalysts, nanofiltration, nanowires, nanotubes, and quantum dots (Figure 1). Other developing countries, such as Thailand, the Philippines, Chile, Argentina, and Mexico, are also pursuing nanotechnology[1].

Figure 1: Quantum Dots for Disease Diagnostics

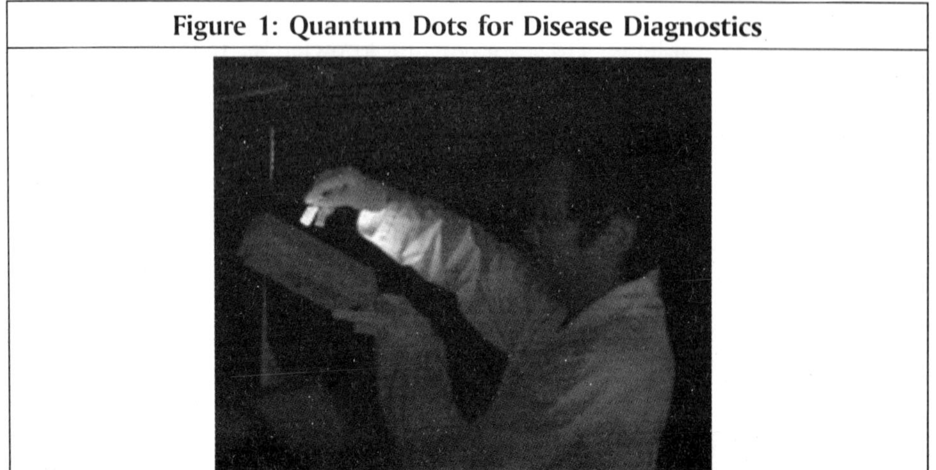

Quantum dots may be used for cheap, efficient **handheld** diagnostic devices available at point-of-care institutions in developing countries.

Science and technology alone are not the answer to sustainable development challenges. Like any other science and technology waves, nanoscience and nanotechnology are not "silver bullets" that will magically solve all the problems of developing countries; the social context of these countries must always be considered. Nevertheless, science and technology are a critical component of development[6]. The 2001 Human Development Report[7] of the UN Development Program clearly illustrates the important roles of science and technology in reducing mortality rates and improving life expectancy in the period 1960-1990, but it did not emphasize nanotechnology specifically. In a report released in early 2005[8], the UN Task Force on Science, Technology and Innovation (part of the process designed to assist UN agencies in achieving the UN MDGs) addresses the potential of nanotechnology for sustainable development.

Top Ten Nanotechnologies Contributing to the MDGs

In order to provide a systematic approach to address sustainable development issues in the developing world, we have identified and ranked the ten applications of nanotechnology most likely to benefit developing countries. We used a modified Delphi Method, as described in our Top Ten Biotechnologies report[9] to identify and prioritize the applications and to achieve consensus among the panelists.

We recruited an international panel of 85 experts in nanotechnology who could provide the informed judgments that this study required, of which 63 completed the project. We selected the panelists based on contacts identified in our previous study on nanotechnology in developing countries[1]. A conscious effort was made to balance the panel with respect to gender, specialty areas within nanotechnology, and geographic distribution. Of the panelists, 38 (60%) were from developing countries and 25 (40%) from developed countries; 51 panelists (81%) were male and 12 (19%) were female.

We posed the following open-ended question: "Which do you think are the nanotechnologies most likely to benefit developing countries in the areas of water, agriculture, nutrition, health, energy, and the environment in the next 10 years?" These areas were identified in the 2002 UN Johannesburg Summit on Sustainable Development[10]. We asked the panelists to answer this question using the following criteria derived from our previous Top Ten Biotechnologies study:

Impact

How much difference will the technology make in improving water, agriculture, nutrition, health, energy, and the environment in developing countries?

Burden

Will it address the most pressing needs?

Appropriateness

Will it be affordable, robust, and adjustable to settings in developing countries, and will it be socially, culturally, and politically acceptable?

Feasibility

Can it realistically be developed and deployed in a time frame of ten years?

Knowledge Gap

Does the technology advance quality of life by creating new knowledge?

Indirect Benefits

Does it address issues such as capacity building and income generation that have indirect, positive effects on developing countries?

Three Delphi rounds were conducted using e-mail messages, faxes, and phone calls. In the first round, the panelists proposed examples of nanotechnologies in response to our study question. We analyzed and organized their answers according to common themes and generated a list of twenty distinct nanotechnology applications. This list was reviewed for face and content validity by two nanotechnologists external to the panel. In the second Delphi round, the panelists ranked their top ten choices from the 20 applications provided and gave reasons for their choices. To analyze the data, we produced a summative point score for each application, ranked the list, and summarized the panelists' reasons. Then we redistributed the top 13 applications, instead of the top ten, to generate a greater number of choices for increased accuracy in the last round. Thus, the highest score possible for an application was 819 (63×13). The final Delphi round was devoted to consolidating consensus by re-ranking the top ten of the 13 choices obtained in the previous round and to gather concrete examples of each application from the panelists.

Our results, shown in Table 1, were compiled from January to July 2004. They display a high degree of consensus with regard to the top four applications: all of the panelists cited at least one of the top four applications in their personal top four rankings, with the majority citing at least three.

Table 1: Correlation between the Top Ten Applications of Nanotechnology for Developing Countries and the UN Millennium Development Goals

Ranking (Score)	Applications of Nanotechnology	Examples	Comparison with the MDGs
1(766)*	Energy storage, production, and conversion	Novel hydrogen storage systems based on carbon nanotubes and other lightweight nanomaterials. Photovoltaic cells and organic light-emitting devices based on quantum dots.	VII
			Contd...

Contd...			
		Carbon nanotubes in composite film coatings for solar cells. Nanocatalysts for hydrogen generation. Hybrid protein-polymer biomimetic membranes.	
2(706)	Agricultural productivity enhancement	Nanoporous zeolites for slow-release and efficient dosage of water and fertilizers for plants, and of nutrients and drugs for livestock. Nanocapsules for herbicide delivery. Nanosensors for soil quality and for plant health monitoring. Nanomagnets for removal of soil contaminants.	I, IV, V, VII
3(682)	Water treatment and remediation	Nanomembranes for water purification, desalination, and detoxification. Nanosensors for the detection of contaminants and pathogens. Nanoporous zeolites, nanoporous polymers, and attapulgite clays for water purification. Magnetic nanoparticles for water treatment and remediation. TiO_2 nanoparticles for the catalytic degradation of water pollutants.	I, IV, V, VII
4(606)	Disease diagnosis and screening	Nanoliter systems (Lab-on-a-chip). Nanosensor arrays based on carbon nanotubes. Quantum dots for disease diagnosis. Magnetic nanoparticles as nanosensors. Antibody-dendrimer conjugates for diagnosis of HIV-1 and cancer. Nanowire and nanobelt nanosensors for disease diagnosis. Nanoparticles as medical image enhancers.	IV, V, VI
5(558)	Drug delivery systems	Nanocapsules, liposomes, dendrimers, buckyballs, nanobiomagnets, and attapulgite clays for slow and sustained drug release systems.	IV, V, VI
			Contd...

Contd...			
6(472)	Food processing and storage	Nanocomposites for plastic film coatings used in food packaging. Antimicrobial nanoemulsions for applications in decontamination of food equipment, packaging, or food. Nanotechnology-based antigen detecting biosensors for identification of pathogen contamination.	I, IV, V
7(410)	Air pollution and remediation	TiO_2 nanoparticle-based photocatalytic degradation of air pollutants in self-cleaning systems. Nanocatalysts for more efficient, cheaper, and better-controlled catalytic converters. Nanosensors for detection of toxic materials and leaks. Gas separation nanodevices.	IV, V, VII
8(366)	Construction	Nanomolecular structures to make asphalt and concrete more robust to water seepage. Heat-resistant nanomaterials to block ultraviolet and infrared radiation. Nanomaterials for cheaper and durable housing, surfaces, coatings, glues, concrete, and heat and light exclusion. Self-cleaning surfaces (e.g., windows, mirrors, toilets) with bioactive coatings.	VII
9(321)	Health monitoring	Nanotubes and nanoparticles for glucose, CO_2, and cholesterol sensors and for in-situ monitoring of homeostasis.	IV, V, VI
10(258)	Vector and pest detection and control	Nanosensors for pest detection. Nanoparticles for new pesticides, insecticides, and insect repellents.	IV, V, VI
*The maximum total score an application could receive was 819.			

To further assess the impact of nanotechnology on sustainable development, we have compared the top ten applications with the UN Millennium Development Goals (Table 1 and Figure 2). The MDGs are eight goals that aim to promote human development and encourage social and economic sustainability[11]. In 2000, all 189 member states of the UN committed to achieve the MDGs by 2015. The MDGs are: (i) Eradicate extreme poverty and hunger; (ii) Achieve universal

Figure 2: Comparison between the Millennium Development Goals and the Nanotechnologies most likely to Benefit Developing Countries in the 2004-2014 Period

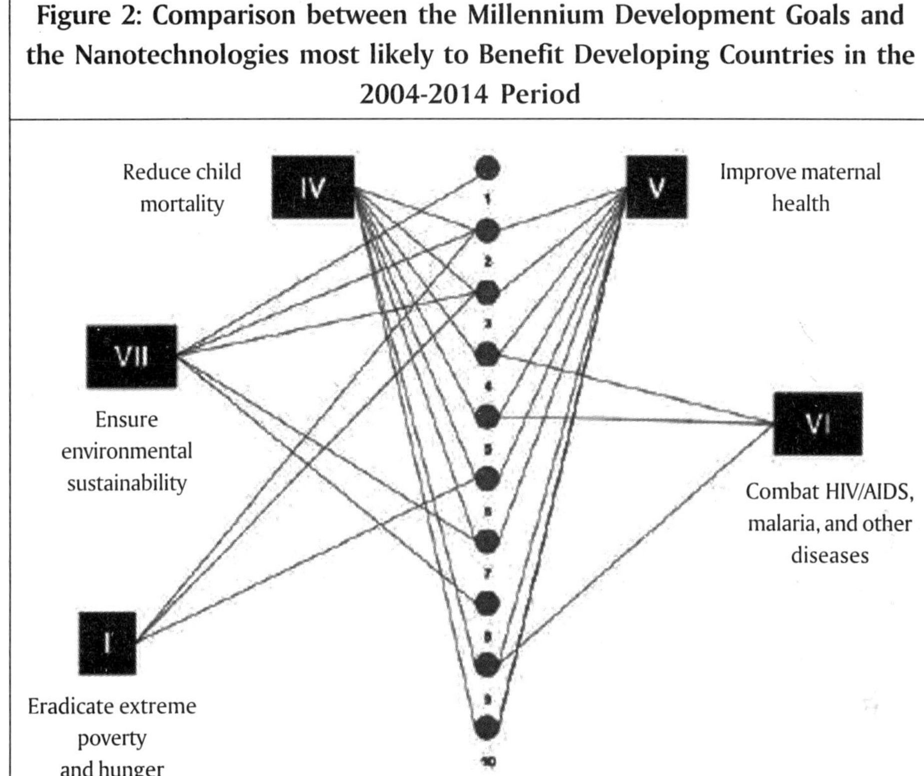

primary education; (iii) Promote gender equality and empower women; (iv) Reduce child mortality; (v) Improve maternal health; (vi) Combat HIV/AIDS, malaria, and other diseases; (vii) Ensure environmental sustainability; and (viii) Develop a global partnership for development. As shown in Table 1 and Figure 2, the top ten nanotechnology applications can contribute in achieving the UN MDGs.

Addressing Global Challenges Using Nanotechnology

What can the international community do to support the application of nanotechnology in developing countries? In 2002, the National Institutes of Health (NIH) conceptualized a roadmap for medical research to identify major opportunities and gaps in biomedical investigations. Nanomedicine is one of the areas of implementation that has been outlined to address this concern. Several of the applications of nanotechnology that we have identified in our study can aid the

NIH in this process by targeting the areas of research that need to be addressed in order to combat some of the serious medical issues facing the developing world.

To expand on this idea, we propose an initiative called "Addressing Global Challenges Using Nanotechnology," to accelerate the use of nanotechnology to address critical sustainable development challenges. We model this proposal on the Foundation for the NIH/Bill and Melinda Gates Foundation's Grand Challenges in Global Health[12], which itself was based on Hilbert's Grand Challenges in Mathematics.

A grand challenge is meant to direct investigators to seek a specific scientific or technological breakthrough that would overcome one or more bottlenecks in an imagined path in solving a significant development problem (or preferably, several)[12]. A scientific board similar to the one created for the Grand Challenges in Global Health, with strong representation of developing countries, will need to be established to provide guidance and oversee the program. The top ten nanotechnology applications identified in Table 1 are a good starting point for defining the grand challenges.

The funding to address global challenges using nanotechnology could come from various sources, including national and international foundations, and from collaboration among nanotechnology initiatives in industrialized and developing countries. These funds could be significantly increased if industrialized nations adopt the target set in February 2004 by Paul Martin, Prime Minister of Canada: that 5% of Canada's research and development investment be used to address developing world challenges[13]. In parallel to the allocation of public funds, policies should provide incentives for the private sector to direct a portion of their research and development toward funding our initiative. The UN Commission on Private Sector and Development report *Unleashing Entrepreneurship: Making Business Work for the Poor*[14] underscores the importance of partnerships with the private sector, especially the domestic private sectors in developing countries, in working to achieve the MDGs.

Perhaps most importantly, our results can provide guidance to the developing countries themselves to help target their growing initiatives in nanotechnology[15]. The goal is to use nanotechnology responsibly[16] to generate real benefits for the 5 billion people in the developing world.

(Fabio Salamanca-Buentello, Deepa L Persad, Erin B Court, Douglas K Martin, Abdallah S Daar and Peter A Singer. All authors are at the University of Toronto Joint Centre for Bioethics (Toronto, Canada) and the Canadian Program on Genomics and Global Health (Toronto, Canada).

Douglas K. Martin is also at the Department of Health Policy, Management and Evaluation, University of Toronto.

Abdallah S. Daar is also at the Department of Public Health Sciences and Surgery, University of Toronto, and the McLaughlin Centre for Molecular Medicine (Toronto, Canada).

Peter A. Singer is also at Department of Medicine, University of Toronto and University Health Network. Abbreviations: NIH, National Institutes of Health; MDGs, Millennium Development Goals, E-mail: peter.singer@utoronto.ca Competing Interests: Peter A. Singer is on the editorial board of PLoS Medicine.

Fabio Salamanca-Buentello can be reached at f.salamanca.buentello@utoronto.ca).

Endnotes

[1] Court E, Daar A S, Martin E, Acharya T, Singer P A (2004) "Will Prince Charles *et al.* diminish the opportunities of developing countries in nanotechnology?" Available: *http://www.nanotechweb.org/articles/society/3/1/1/1.* Accessed February 21, 2005.

[2] "US, Indian high technology will benefit through cooperation". (2003) Available: *http://newdelhi.usembassy.gov/wwwhpr0812a.html.* Accessed January 27, 2005.

[3] Bapsy P P, Raghunadharao D, Majumdar A, Ganguly S, Roy A, *et al.* (2004) DO/NDR/02 "A novel polymeric nanoparticle paclitaxel: Results of a phase 1 dose escalation study". J Clin Oncol 22 14S:2026. Find this article online.

[4] [Anonymous]. (2003) "China's nanotechnology patent applications rank third in world". Available: *http://www.investorideas.com/Companies/Nanotechnology/Articles/China's Nanotechnology1003,03.asp.* Accessed January 27, 2005.

[5] Meridian Institute. (2004) *Report of the international dialogue on responsible research and development of nanotechnology*. Attachment F Available: *http://www.nanoandthepoor.org/Attachment_F_Responses_and_Background_Info_040812.pdf.* Accessed February 21, 2005.

[6] Sachs J (2002) "The essential ingredient". *New Sci* 2356:175. Find this article online.

[7] UN Development Programme. (2001) *Human Development Report.* Available: *http://www.undp.org/hdr2001/completenew.pdf.* Accessed February 21, 2005.

[8] UN Millennium Project 2005. (2005) "Innovation: Applying knowledge in development. Task force on science, technology and innovation". Available: *http://unmp.forumone.com/eng_task_force/ScienceEbook.pdf.* Accessed February 21, 2005.

[9] Daar A S, Thorsteinsdóttir H, Martin D, Smith A C, Nast S, *et al.* (2002) "Top ten biotechnologies for improving health in developing countries". Nat Genet 23:229–232. Find this article online.

[10] World Summit on Sustainable Development. (2002) Available: *http://www.johannesburgsummit.org/html/documents/wehab_papers.html.* Accessed January 27, 2005.

[11] United Nations. (2000) "UN millennium development goals". Available: *http://www.un.org/millenniumgoals/.* Accessed January 27, 2005.

[12] Varmus H, Klausner R, Zerhouni E, Acharya T, Daar AS, *et al.* (2003) "Grand challenges in global health". *Science* 302:398–399. Find this article online.

[13] Government of Canada, Office of the Prime Minister. (2004) Complete text and videos of the Prime Minister's reply to the speech from the throne. Available: *http://www.pm.gc.ca/eng/news.asp?id=277.* Accessed January 27, 2005.

[14] UN Development Programme, Commission on the Private Sector and Development. (2004) "Unleashing entrepreneurship: Making business work for the poor". Available: *http://www.undp.org/cpsd/indexF.html.* Accessed February 21, 2005.

[15] Meridian Institute. (2005) "Nanotechnology and the poor: Opportunities and risks". Available: *http://www.nanoandthepoor.org/gdnp.php.* Accessed February 21, 2005.

[16] The Royal Society and The Royal Academy of Engineering. (2004) "Nanoscience and nanotechnologies: Opportunities and uncertainties". Available: *http://www.royalsoc.ac.uk/policy.* Accessed January 27, 2005.

Progress Of Nanotechnology In The Developing World

– Anil Varma

In developing nations diseases like TB are rampant. Nanotechnology helps to go a long way in the fight against such diseases. In nations like India research in nanotechnology has enabled the development of diagnostic kits which are uncomplicated, precise, small and secure. These kits have carbon based nanotube filters that eliminate germs of nano scale also. The main benefit of this is that these kits are cheap and affordable which is the main benefit for the people in developing nations. A typhoid detection kit has also been developed by making use of nanosensors so that the regular Widal tests can be taken with more definite results. In countries like India an important initiative has been taken where methods have been developed so as to create cabon nanotube filters which locate and destroy the nano scale impurities that are existent in water. Special nanotube filters have been fabricated which make it possible to eliminate viruses of 25 nanometer size from the water. This is particularly helpful in developing countries where clean and pure water is still a distant dream in many areas.

Nanotechnology has been given special attention because of its exclusive potential to ease the misery of the poor in the developing nations. On a long term basis nanosystems that are productive will ensure the cheaper and better organized processes of production which would lower the price of all wares. Traditionally technology has been the driver behind economic progress and development. The opening of nanotechnology to the developing world will have overpoweringly positive effects on the potential for growth of these nations.

(Anil Varma, Consulting Editor, The Icfai Research Centre Pune. He can be reached at avarma@iupindia.org.)

14

Micro Nano Technology Commercialisation Pitfalls

Henne van Heeren, Patric Salomon,
Lia Paschalidou and Ayman El-Fatatry

Ever since the realisation that MNT (Micro Nano Technology) offers potential opportunities for accessing new markets and expanding existing ones, the race for exploitation began. These opportunities encompassed technical, financial as well as enabling capabilities. As a result, successful ventures were initiated by (large) companies, institutes and entrepreneurs. These have charted the way forward. Failed adventures have, however, provided sufficient warnings of the pitfalls and challenges encountered along the way to commercialisation. This "white paper[1]" attempts to summarise some of the more obvious problems and other less clear hurdles likely to be encountered by prospective new business ventures.

[1] EnablingMNT publishes briefs entitled "white papers" addressing important issues of relevance to MNT commercialisation.

The Commercialisation Lifecycle

Lifecycles depicting the commercialisation of all products, in general, tend to follow similar paths from the concept/idea stage through *commercialisation* and market penetration.

Generally speaking, such a lifecycle process for commercialisation is as shown below:

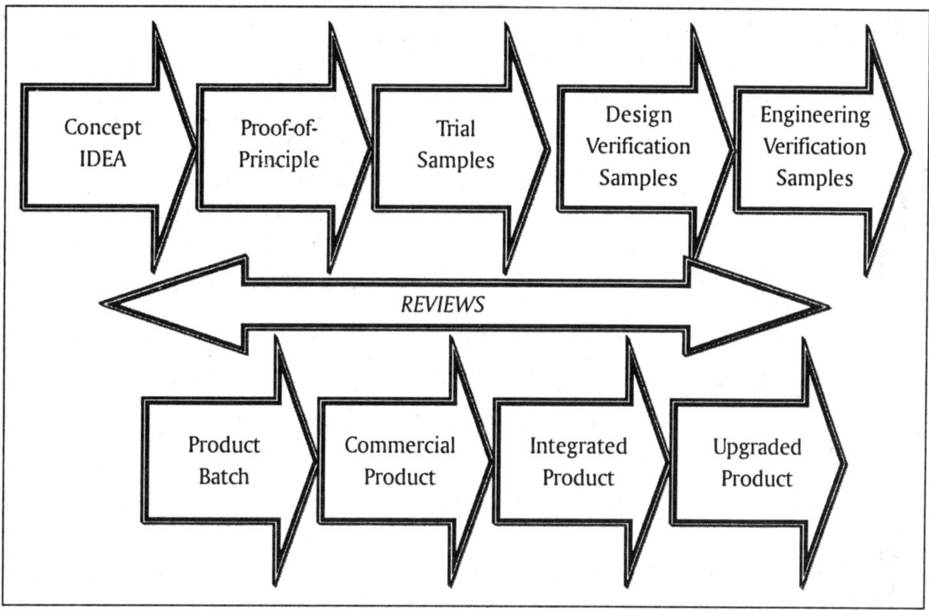

Within such a framework, a concept or idea migrates towards commercialisation via an iterative process of design verification and optimisation. This process tends to be underpinned by marketing information and customer involvement.

For microsystems, the complexity and the multi-disciplinary nature of the product in its own right tend to add complications which necessitate an understanding of the various factors likely to affect progress from one step to the next. These factors interact at all levels and throughout the lifecycle pathway as shown below:

Each phase from idea to full commercialisation tends to require specific capabilities and therefore its own specific set of stakeholders and other participants.

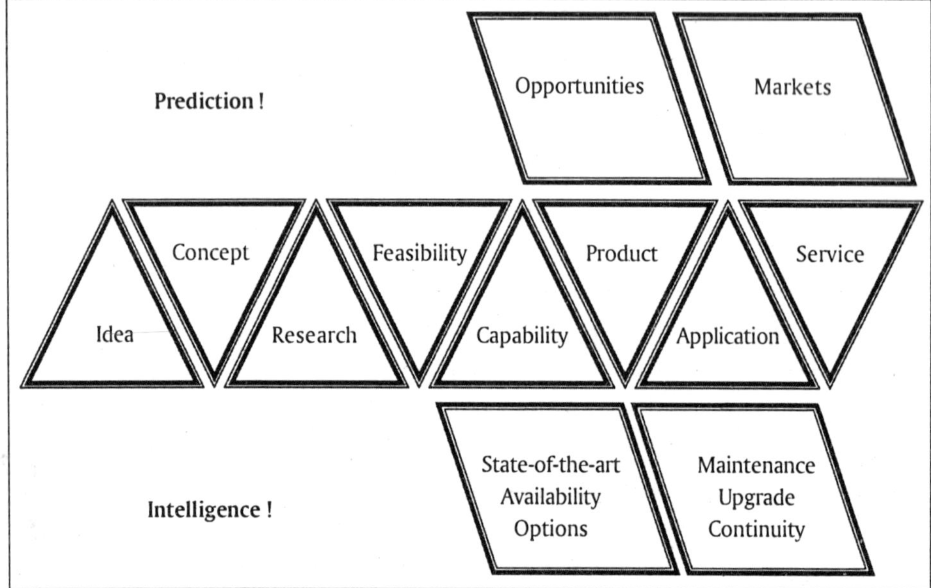

The interrelationships between the various players in the MNT industrialisation process tend to be further complicated by a relatively limited experience in this industry vis-à-vis what is the case for more mature industries such as microelectronics. Of note is the, as yet, undeveloped need to set up and maintain a process for post commercialisation, which will address maintenance and upgradeability.

Furthermore, the multi-disciplinary nature of the various technologies (and know-how) required tends to complicate the process of optimisation, as certain disciplines may remain, untapped or are adopted at a late stage in the process.

In this brief note, a number of more major issues are highlighted as potential pitfalls likely to be encountered along the path of commercialisation.

1. Being Part of a Chain

A Collaborative Venture?

The essence of microsystems lies in the multi-disciplinary nature of the underpinning technologies. The integration of mechanical engineering with that of semiconductor processing, as well as electronics, is but one example that typifies the development of a product such as the MNT gyro or accelerometer. Other disciplines tend to be drawn along with the development lifecycle as and when appropriate. Although this approach tends to be normal for most stand-alone products, the multi-functional potential of microsystems when fully exploited will undoubtedly necessitate closer interactions between the disciplines upfront (during the concept assessment stage). Basically, ventures hoping to commercialise concepts must be aware of the future potential of their devices within the context of larger systems and more complex environments.

In this context, all parties likely to benefit from such concepts must be included at the appropriate moment during the lifecycle. This issue is illustrated to some extent by the following diagram where disciplines such as wafer-scale designers are engaged alongside packaging experts as well as systems integrators.

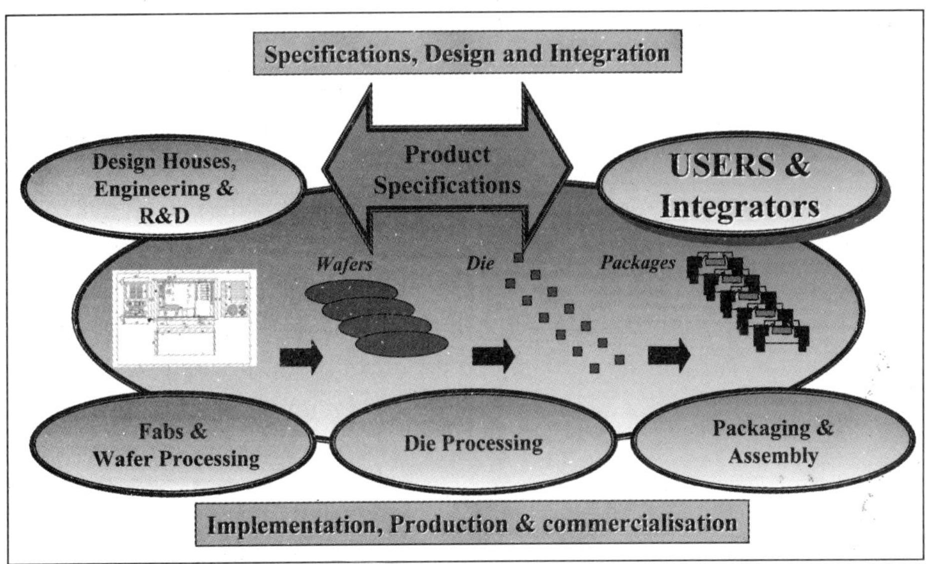

Or a Partnership?

The unique nature of microsystems both as a new entrant into the technological sphere as well as its multi-disciplinary characteristics renders all aspects relating to its commercialisation of particular interest to a wide range of stakeholders. In essence, the variously disciplined technologists have to adapt a new mode of collaborative integration. Likewise, the financial backers and investors have to understand the evolutionary nature of the resultant products (increasing in functionality and intelligence). Equally pertinent are all the other stakeholders responsible for realising the commercial outcome, including the fabrication and design houses, the test and qualification engineers and the end-users. The latter group of stakeholders are increasingly important for the success of the MNT products due to the potential on offer in terms of ambient intelligence and miniaturisation. These diverse expectations (from disparate cultures) need to be understood and contained within realistic boundaries.

Lack of Infrastructure

The successful commercialisation of products is highly dependent on the availability of the right infrastructure at the right time. This, rather obvious, statement is particularly relevant to microsystems because of the diversity of the infrastructure required for the realisation and delivery of a marketable end product. Moreover, the diverse infrastructure tends to be capital intensive demanding high throughput and maximum utilisation for economic returns – both being highly stretching demands for MNT products within the currently known market size (with a few notable exceptions).

As a result, many ventures have been established via one of the two distinct routes: (1) As an offshoot from an academic/research establishment or (2) as an offshoot from a large industrial concern. Both options utilise equipment that is capitalised across a large number of applications, which include the MNT development and research within it. Approaches relying on the use of product-specific and exclusive infrastructure tend to face intolerable pressures from investors for returns (mostly ending with disaster) or end up targeting in the, less lucrative, niche markets.

Successful ventures tend to include all players within the commercialisation plans, ensuring that equipment availability (during commercialisation and subsequent upgrades) is pivotal to the plans. Equally relevant are, in some cases, plans for fall-back scenarios to redress the dependence on somewhat less stable service providers.

The fact that MNT products are likely to require piece-meal access to fabrication processes and facilities is a problem faced by many commercialisation ventures. In such a situation process and yield optimisation are often neglected as more urgent (and initially more profitable) alternative activities take precedence. Therefore, the transfer of the technology and the processes from a research environment to that of industry tends to be incomplete. The result is a process with products that tend to be characterised with performance drifts and bulk quantity limitations. Once again, the large ventures tend to succeed on this front given their ability to withstand initial losses and large investments.

2. Quality

Standards – What Standards?

The microsystems industry is, in all aspects, rather embryonic when compared to the better-established industries of microelectronics, IT amongst others. In addition, as emphasised above, the multi-disciplinary nature of MNT is specific to the characteristics of this novel technology. Furthermore, the technologies and products are developed to address a multi-market scenario. These characteristics, particular to MNT, make it very difficult for standards to be set and followed within this industry. This applies equally to product performance standards and to fabrication standards. The resultant scarcity of clear standards allows operators to produce solutions (and products) which generally do not conform to wider system requirements but dictate their own, and to slow integration, interaction and maintenance. Clearly, organisations with influence and/or market monopoly tend to enforce their preferred solutions, which may not be optimum for all users/buyers. Equally problematic are the small ventures, which attempt to penetrate protected markets and an established way-of-doing-things.

The lack of fabrication standards is also a major issue worth noting. Processes used for MNT may not fall within the realm of the well-established ones (particularly in the case of silicon processing). In addition, as MNT products are unlikely to require full-time utilisation of the production equipment, the development will be consigned to a secondary priority relative to the main product of a fabrication line. Also, to meet product requirements and satisfy customers, fabrication standards may not be followed, resulting in inconsistencies in quality. This situation is difficult to control, particularly whilst MNT markets remain uncertain.

3. Handling of IP

Exclusivity – Wide Dissemination Versus Protective Strategies

The generic nature of micro and nano technologies with regard to its applicability across a wide range of systems offers a conflict of interest between the desire to disseminate intellectual property (IP) (to maximise exploitation) and the importance of protecting the self same IP (to maintain a competitive edge). Furthermore, new ventures tend to rely heavily on outsourcing and the use of facilities on an ad hoc basis. Protecting IP within such shared environments tends to be complicated and the approach taken is one of minimal dissemination. The disadvantage of such an approach lies in the low levels of interactions between (potential) stakeholders and a sluggish technology transfer mechanism not conducive to optimise yield and improve processes. In addition, such shared environments provide opportunities for takeover and control (positive or negative).

Patenting before Commercialisation

The appeal for researchers (and budding entrepreneurs) to patent as soon as an idea is close to maturity is understandable. This situation does, however, open up potential problems that are often underestimated. The most notable of such problems is disclosure; more often than not, ideas are patented at an early stage of development enabling competitors to "upgrade and circumvent" the patent applications. Other dangers include buy-out by competitors or tedious battles on infringement. In either case the costs tend to be prohibitive for the small ventures.

The alternative is for ventures to patent after maturity hence delaying the decision at the risk of losing ownership; this tactic is often adopted by companies wishing to promote uptake and encourage standardisation. Early return on investment for such cases is difficult and licensing of processes almost impossible.

4. Narrow Vision Pitfalls

A Disruptive Technology?

Microtechnology is, generally, described as both an enabling technology as well as a disruptive technology capable of displacing well-established products and methodologies.

This differentiation between what could potentially expand markets and what may outplace competitors is very important. It is the threat of the new as much as the ignorance of the new that may stifle the concept from the start. Ironically, many disruptive technologies (or solutions) tend to be identified as such by chance. Only the keen observer (with foresight) is likely to be the first to grasp such chances.

Challenges associated with disruption are likely to affect large established organisations more than the newer and smaller activities, as the latter are less entranced in well proven methods and, as a rule, more flexible compared to larger old establishments.

Application is Leading

Applications-pull based products tend to succeed commercially for obvious reasons whilst technology-push products, generally, face several barriers to enter new territories. Microsystems-based products tend to fall firmly within the latter category of products. This is not entirely due to the novelty of the technology associated with MNT but also due to the new applications enabled by it. In essence, the multifunctional nature of the products enable new and/or more complex applications such as lab-on-a-chip. Such (sub)systems have opened up new opportunities which were previously either unknown or impossible.

This fact does pose a challenge to new ventures in MNT, whereby, the systems integrator (end-user) is neither aware nor prepared for the risk of introducing new, multi-functional, capabilities. Unlike most technology-push products, MNT complicates the challenge further by charting new ground into what could be well established and/or conservative application domains, such as medical, pharmaceutical and automotive. In other words, MNT provides opportunities that may well have been undiscovered. This makes justification more difficult and plans (such as technology roadmaps) less obvious and clear to draw.

Finally, well-established applications (particularly in the automotive sector) tend to be highly risk-averse. A challenge that is most difficult to overcome by MNT advocates due, mainly, to cost, reliability and availability issues.

System Approach

Two aspects characterise microsystems, namely; the multi-disciplinary nature of the underlying technologies and the level of integration required to develop multi-functional devices. It is either, or both, of these characteristics which place microsystems in a unique position from the perspective of an overall systems design and/or its evolution as well as the intended application.

As a result microsystems are enabling an increasing complexity of capability, functionality and design. In essence, microsystems technology is advancing towards the realisation of, multi-functional, intelligent devices such as the lab-on-a-chip for diagnostics or miniaturised inertial measurement units for aerospace/automotive applications. These sub-systems will, in future, become integrated within larger systems to form a part of an intelligent, ambient environment populated with sensors, actuators and wireless transceivers. In such systems, intelligence will be distributed and shared *via, ad-hoc,* wireless networks and data/knowledge will be processed in real-time.

It is vital for the technologists, the end-users and the strategists to recognise this capability at an early stage in order to fully exploit this potential via evolutionary design concepts and philosophies.

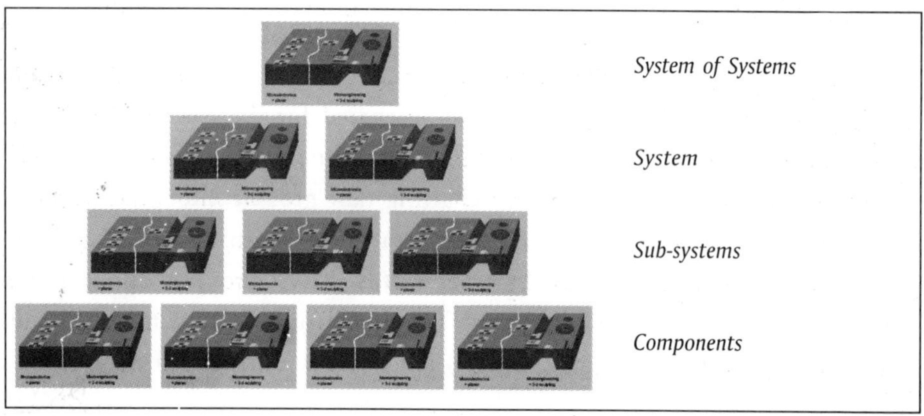

The above diagram illustrates the potential hierarchy of systems evolution in MNT.

The "Not Invented here" Syndrome

Ownership of concepts and ideas is considered vital to the success of a business venture. In this context, companies strive to protect their intellectual property via patents, design rules and/or simple secrecy. Investment associated with such protection tends to be high and research programmes initiated to prove developments absorb considerable funds that investors hope will provide significant returns within short timescales. Once entrenched within such programmes, companies and decision makers are, generally, reluctant to reconsider their plans in the face of new ideas/solutions/options. The "not invented here" ideology predominates and investment may well be ill-placed.

However, technologists involved with MNT have recognised that a multi-disciplinary team is essential for the successful commercialisation of a product. Such teams will encompass mechanical, electronic, optical design and possibly bio-chemical engineers as well as packaging experts, consultants and the end-users. This resulted in a shift from the "not invented here" syndrome to the "supply chain" philosophy.

(un)Realistic Estimations of the Target/Potential Market(s)

Market sizes for MNT products are difficult to estimate, mainly due to the varied use of definitions for this technology. In essence, a market size for an MNT based product can be highly exaggerated if it is estimated on the basis of a sub-system level as opposed to that of the underpinning component; print-head versus printer, micro-mirror versus projector; heart pace maker versus accelerometer are but some examples where the estimates could well be overblown. Conversely, the market size may be greatly under-estimated as is the case where MNT is an enabling (but vital) component within the overall (sub)system; fibre alignment v-grooves and micromachined passive RF components are typical examples for this case.

Furthermore, the markets predicted by several study groups (e.g., NEXUS), although large in total (around $25Bn by 2009), they are relatively small from the viewpoint of well established industries such as that of IT or microelectronics. This perception tends to keep the "big" players away thereby perpetuating the situation of small and niche. Of late, however, some of the larger organizations have begun to increase their involvement in MNT, particularly for products aimed at the lucrative telecommunications and medical markets.

The following diagram illustrates this latter point with regard to the entrance of equipment suppliers within the domain (playing field) of MNT[2]:

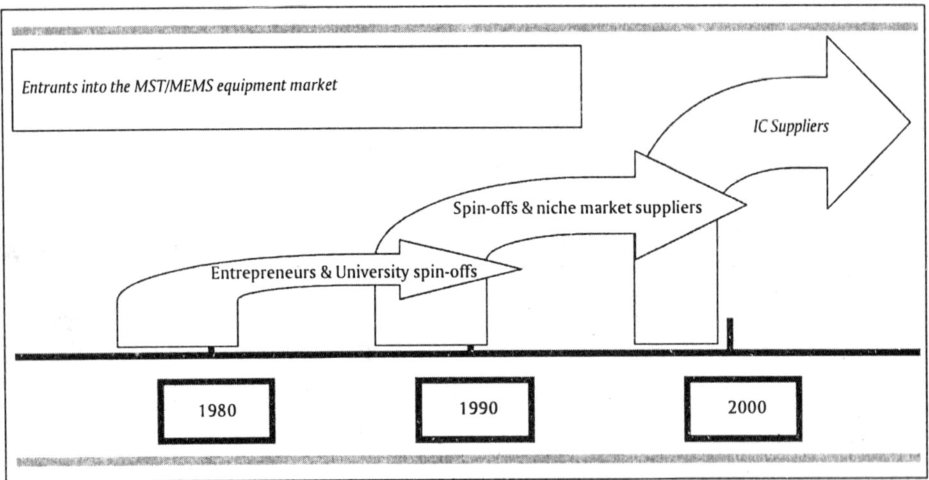

5. Concluding Remarks

As with almost all new ventures and adventures, there are many potential pitfalls and challenges, which need to be overcome. The ventures likely to succeed are those which have learnt from their (or others') mistakes. They are also those who are prepared and have the foresight to assess risks and mitigate those risks with both prior knowledge and preparedness.

Although most of the foregoing remarks seem cautionary and, perhaps, over emphasized, the aim of this brief note is to present the broader picture. In essence, this brief should help uncover potential problems likely to befall new comers to this field of technology.

Finally, in a time marked by many examples of paradigm shifts, such as that created by the micro, and even more so, the nano technology, new ventures get new opportunities and a chance of entering new markets.

2 See also the enablingMNT reviews of Front-End, Back-End and Nanotechnology production equipment, *www.enablingmnt.com*

For More Information

The enablingMNT Team have the experience and the expertise to help in understanding and addressing many of the above listed challenges. In this context, enablingMNT provides in-depth reports and assessments on technology related issues associated with MNT commercialisation (see also *www.enablingmnt.com*).

Call or email us at:

Henne van Heeren	Patric Salomon
E-mail: Henne@enablingmnt.com	E-mail: Patric@enablingmnt.com
Tel.: +31 654 954 621	Tel.: +49 302 435 7870
EnablingM3	4M2C PATRIC SALOMON GmbH
Elzenlaan 154	Cranachstrasse 1
3319 VC Dordrecht	12157 Berlin
The Nethmerlands	Germany

About EnablingMNT

EnablingMNT is the brand adopted by an international team of experts with many years of experience in the business of micro nano technologies (MNT), namely: Henne van Heeren, Patric Salomon, Dr. Lia Paschalidou, and Dr. Ayman El-Fatatry. With their track record in technology development, engineering, systems integration, manufacturing, project management, business development, market analysis, and strategy consulting, the enablingMNT team is well versed in all commercialization issues relating to the supply chain of MNT: from concept through to product. With this expertise, the team addresses specific customer requirements, particularly, in relation to market intelligence, technical insight, strategic development plans and investment decisions. Based in the UK, Germany and the Netherlands with excellent contacts in the Far East and the USA, the team is able to analyse the worldwide activities in the MNT industry. This collaborative research of the *enablingMNT team* has, to date, culminated in a series of in-depth intelligence reviews on the supply chain of micro and nano technologies (MNT): *enablingMNT Industry Reviews.*

About EnablingMNT Industry Reviews

Driven by user/supplier demands for technology-pull, the reviews in this *enablingMNT series* are focused on all aspects of the micro and nanotechnology

(MNT) product lifecycle; from concept through to production. They describe how a selected topic is linked into the supply chain from product idea, technology development, engineering, prototyping, manufacturing, packaging and test, into volume production. The *enablingMNT reviews* analyse the critical issues encountered during the creation process of MNT-based products and provide guidance with regards to the approaches taken to overcome barriers and challenges. All reviews include a comprehensive listing of services offered in the field (including current contact information). The emphasis is on the provision and availability of fully commercial services, infrastructure, and suppliers considered to be essential for the commercial realisation of MNT-based products. It is expected that *enablingMNT industry reviews* will motivate new applications for MNT, enhance confidence in the technology as well as help identify funding sources for research and development. They highlight trends from the facts and the prospects. Supported by significant players in the field, the reviews are produced and edited by the *enablingMNT* team. *enablingMNT industry reviews* are offered at a competitive price of €280 per specialist report.

The reviews available, to date are (priced at €280 each):

- Foundries for MST/MEMS (NEW!)
- Suppliers of Processing Services for MST/MEMS
- Test- and Measurement Equipment Suppliers for MST/MEMS
- Suppliers of Materials for MST/MEMS Production
- MNT Web Directories and On-line Communication Channels
- Equipment Suppliers for Nanofabrication
- Back-End Manufacturing Equipment Suppliers for MST/MEMS
- Front-End Manufacturing Equipment Suppliers for MST/MEMS
- Packaging&Assembly Providers for MST/MEMS
- Design&Engineering Companies for MST/MEMS

Coming soon:

- Design & Engineering Companies for MST/MEMS 2006 update
- MNT Design, Modelling, Simulation Tools
- Packaging & Assembly Providers for MST/MEMS 2006 update

To order a copy and/or for more information, please contact the publisher Patric Salomon at:

4M2C Patric Salomon GmbH, Cranachstrasse 1, 12157 Berlin, Germany

Phone: +49 30 2435 7870,

E-mail: *info@enablingMNT.com*

www.enablingMNT.com

The enablingMNT team acknowledge the help of the Industry Reviews sponsors:

WTC - Wicht Technology Consulting, Germany *www.wtc-consult.de*

MST News, Germany *www.mstnews.de*

SUSS Microtec, Germany *www.suss.com*

NEXUS, Europe *www.nexus-mems.com*

MEMS Industry Group, USA *www.memsindustrygroup.org*

SEMI/IMSG, Europe *www.semi.org*

enablingMNT, Europe *www.enablingMNT.com*

MANCEF, USA *www.mancef.org*

(Henne van Heeren (NL) runs his own company, EnablingM3, which offers consultancy and market research services in the field of MNT. Patric Salomon (D) runs his own consultancy company 4M2C PATRIC SALOMON GmbH, with a focus on all aspects of marketing and strategy support for the MNT dependant industry. Dr. Lia Paschalidou (UK) is an independent market intelligence consultant, providing B2B market research and analysis services to the MNT and related industries. Dr. Ayman El-Fatatry (UK) is responsible for marketing, communication, publicity and business development for the Systems Engineering Innovation Centre, a venture between the BAE SYSTEMS Advanced Technology Centre and Loughborough University. The authors can be reached at Henne@enablingMNT.com, Patric@enablingMNT.com, Lia@enablingMNT.com, A.El-Fatatry@lboro.ac.uk, respectively).

15

Will Nanotechnology Help Developing Economies

Anil Varma

This article takes a look at issues in nanotechnology, specific to developing countries. A very relevant question it raises is "Will nanotechnology offer solutions to the problems faced by developing countries?" as a sequel to the previous question – "Will the new technology complement or substitute existing technologies for the similar problems?" If the technology is a substitute, then it raises the issue of possible adverse impact on employment. Also an important issue is access to developing countries in light of intellectual property rights. If this is not granted, then it would mean most of the promises and advantages will be available to a select few countries in the world.

What is Nanotechnology

Just think about it – Conceptualize that you have the ability to change the arrangement of atoms any which way that you want to. The properties of products that are manufactured are dependant on the way in which those atoms are organized. If the atoms in coal are reorganized, the outcome that we get may be diamonds. If the atoms in sand are reorganized and a minor amount of impurities added to it then, the result can be electronic chips. Also if the atoms in air, soil and water are reorganized then we might get grass.

Nanotechnology is the development of functional components, compositions, mechanisms and techniques which is achieved by manipulating the shape as well as the dimension of matter on a scale, that is very miniscule. The constant pursuit in support of miniaturization has got us into such a position that we are slowly becoming capable of performing acts that so far were being done only by nature, such as creating switches with the use of a single atom, or the production of original proteins.

Why Nanotechnology

The question being asked is that why is the general public talking more and more about nanotechnology, as the 'fantastic & futuristic revolution' when in actuality nanotechnology has existed for quite some time? One of the prime reasons can be that barely in the past couple of decades have we actually had the experimental resources to carry out work, which had a focus on nanoscale activities. The progress in nanotechnology has been corresponded with the innovation of materials like nanotubes etc., and in the recent past, it has been encouraged by a huge amount of fundings from the Government in countries such as India, China and Japan.

Another reason is that the fundamental aim of nanotechnology is to create and produce with definitive accuracy on the atomic dimension with a 'bottom-up' approach. This indicates that, instead of using the traditional method of manufacturing wherein larger size materials are reduced in size, nanotechnology strives to create procedures beginning with the self-assembly of distinct atoms into specific compositions.

Besides the above, the most important reason is the acknowledgment of nanotechnology as an upcoming field, which requires and gives rise to new degrees of multi-disciplinary association and cross-fertilization between the disciplines.

The ever-increasing aspiration and need to categorize technology, consequential to manipulation at a nanoscale level and the progressive amalgamation of technical disciplines at a merging length-scale, has shown the way to the acknowledged term 'nanotechnology', under the realms of which innovative research is increasing and existing research is frequently reorganized.

Impact of Nanotechnology and Developing Countries

In a study that was conducted recently the applications of nanotechnology were categorized as per their prospective advantage for developing countries. Treatment of water, diagnosis of diseases and drug distribution systems were rated between the third and the fifth positions respectively. They were positioned after enhancement of agricultural productivity, which was second, and storage, production and conversion of energy, which was in the first position.

The nature of the impact of nanotechnology on a global level to a great extent be dependant on the answers to the simple but significant questions, which or who followed by what, then where, when and lastly how? Nations, which are developing, will be subjected to contradictory forms of commitment with nanotechnology, but will it be possible to comment on any general bearings? Is it possible that nanotechnology will turn out to be a lucrative industry for the developing countries or on the other hand, will nanotechnology make use of the developing countries and put in jeopardy the markets of these countries, especially in relation to the principal areas of production like rubber, minerals and cotton. Or is it possible that the developing nations will function as the manufacturing hub for the innovations in nanotechnology. Today countries like South Africa and Malaysia are promoted as nations, which have a relative manufacturing advantage with relation to the field of nanotechnology.

Nanotechnology can be channeled and be utilized to tackle a few of the world's most significant problems of development. On the other hand the applications of nanotechnology, which are directed towards the trials faced by the billions of people of the developing world, have not been given organized and methodical prioritization.

Developing Countries Innovate in Nanotechnology

One of the most passionately discussed subjects, and one of the most difficult to determine in advance in the developing dialogue on nanotechnology, is the possible effects it will have on developing nations and underprivileged sections of the populace. There are points of view which are optimistic, in which nanotechnology is regarded to be a universal remedy, and there are points of view which are pessimistic which presume that the gap between the richer section and the poorer section of the populace will increase as a result of the dissemination of this type of technology. The deliberation on these different points of view, with the support of conjectural debates and historical data, is elementary for arriving at a unbiased perspective of the state of affairs.

Many developing nations have initiated projects of nanotechnology so as to strengthen their capabilities and maintain economic development. Nations like India have planned to invest about 20 million dollars between the period 2004 to 2009 for their initiative in the science and technology of nanomaterials. Some private firms are conducting research on drug delivery and also taking part in clinical trials of the delivery of nanoparticles of anti-cancer drugs. The number of applications for patents from China makes it the third country in the ranking after USA and Japan. Even nations like Brazil have projected about 25 million dollars as their budget for the period 2004 to 2007 for research on nanoscience and they have three institutions, four associations and about three hundred scientists who are researching on nanotechnology. The nanotechnology initiative in South Africa resulted in a national association of intellectual researchers involved in the various divisions of nanotechnology. Many other developing nations like Thailand, Philippines, Argentina, Chile and Mexico are also investing into the research of nanotechnology.

The answer to the challenges of sustainable development does not lie only with science and technology. Like the bearing of any other science and technology, nanotechnology and nanoscience are not the "bolt of lightning" that will enchantingly resolve all the issues of developing nations. The social perspective of these nations must always be taken into consideration. All the same, science and technology are a vital part of development.

Many developing nations depend to a great extent on commodities as a main resource of employment and income for their populace as well as earnings from exports. The applications of Nanotechnology which are in the process of being developed could have an influence on the global markets for farming, mineral and supplementary non-fuel merchandise. These applications of nanotechnology could, in some cases, decrease the requirement of some commodities and in some other cases craft fresh or broader markets for commodities. These alterations could have, prospectively across the board and been extensive, socio-economic as well as resulted in further consequences which may be positive or negative for the developing nations. About 70% of the developing nations obtain a minimum of 50% of their export income from commodities. It is estimated that almost one third of the global population is in employment for the production of commodities, with almost 50% of that population employed in the agricultural sector. Familiarity and understanding of the prospective implications of nanotechnology for developing countries which are dependant on commodities and for developing countries universally, is of supreme significance to ensur that the prospects of nanotechnology for these nations are maximized, while its threats are minimized.

Nanotechnology has been acknowledged as an area with potential of technological progression for developing nations in general, because it may facilitate inclusion of more proficient, effectual, and economical materials, goods and procedures, including processes of manufacturing that have humble investment, land, labor, power, and material necessities. Nanotechnology might also produce fresh or larger markets for commodities manufactured by developing countries and also create prospects to manufacture commodity products which include some value addition. Apprehensions have also been brought up that the same distinctiveness that makes nanotechnology promising for developing nations also creates the likelihood that it may put out of place commodities, labour, and businesses and weaken the overall position of developing nations. Some associations are also voiced the apprehension that nanoscale materials could cause unidentified hazards, including threat to the health of humans and threat to the environment, which may be predominantly difficult to classify and administer.

Significant Questions and Issues for Contemplation

Quite a few traversing and intersecting issues and queries should be taken into consideration while assessing and formulating more knowledgeable conclusions in relation to the prospective consequences of applications of nanotechnology for particular economies and sectors. Despite the fact that these issues may be commonly relevant to technologies, the distinctive characteristics of nanotechnology may perhaps result in diverse considerations for every traversing and intersecting issue, which, in turn, may well call for fresh and diverse strategies for tackling these issues. These issues and questions comprise of, but are not essentially restricted to:

- Research and development of products
  - At what juncture is the development of the technology (e.g., available in the market, tested in the field, tested in the lab, or in the early stages of research)?
  - What would be required to progress the technology from the laboratory to the market?
  - Which type of industries does the technology support?
  - Is the research and development designed to meet the needs of development?
  - What should be the incentives, if required, which should be made available to promote dependable research and development, in addition to the acceptance and alteration of new technologies?
- Ecological, human health, and security threats
  - What are the technology's probable ecological, human health, and security threats?
  - How comprehensively have the ecological, human health and security threats been assessed and how can relevant information be retrieved?
  - What is the requirement for studies on ecological, human health, and security threats and the requirement for development of approaches for risk management?
- Social and economic concerns
  - What are the nation's general improvement requirements and strategies and how is it possible for the technology to play a part?
  - Will the technology operate as a complement or alternate for existing wares?
  - How operational are the manufacturers and markets of commodities to envisage and fine-tune to changes in the market triggered by technology?

How Nanoechnology helps Developing Nations

Although there are quite a few prospective applications for nanotechnology it would be difficult to talk about them all explicitly. Some of the applications of nanotechnology are in the fields of consumer products, environment and energy, applications in medicine, electronics, chemicals, materials, the manufacturing industry and the defense sector. Over and above, the benefits and uses in the above mentioned areas, there are nanotechnology applications that will particularly facilitate developing nations. The first area in which applications of nanotechnology can help developing nation is in the field of energy. Nanotechnology facilitates more effective solar cells, hydrogen fuel cells, and additional systems of energy storage which help to decrease the reliance on petroleum and other fuels that are high polluting agents. The next is in the field of agriculture. In this, nanotechnology enables the more competent delivery of the required nutritional elements and herbicides, nanosensors are used to observe and examine the condition of the soil and this technology is also used to remove the contaminants in the soil. In water treatment nanotechnology is used a low cost and convenient method of cleaning and purifying water. In the medical field nanotechnology is used so as to enable us to have cheap disease diagnosis procedures and also enables cheap enhancement of medical images. In the systems of drug delivery nanotechnology develops and improves slow, specific and constant delivery of drugs. Nanotechnology enables the recognition of contamination as well as efficient decontamination which enables efficient food processing and storage. Nanotechnology is used to create more efficient and economical catalytic converters which help in the detection and destruction of air pollutants. Nanotechnology is also used to develop cheap and long lasting materials like surfaces which are self cleaning and concrete or asphalt which is water resistant which helps in developing cheaper and better housing. Nanotechnology is also used effectively in monitoring of health as well as for effective pest control. Over and above the advantages that will arise from the direct application of nanotechnology, there are several other outcomes that will benefit the societies of the developing nations. The speedy development of nanotechnology will result in a very significant growth in the number of jobs. If we compare it to the Information Technology sector then for each employee in the IT sector there will be twice the amount of jobs created in the related non-technical areas. Another probable benefit is that nanotechnology creates a link between the scientific disciplines. Nanotechnology, because of its basic scientific nature will have impact on the various sciences like physics, chemistry, engineering, environmental sciences, biology etc. This will increase the exchange of ideas between sciences and bring deviating sciences together. This will allow for a better amalgamation of scientific principles into education.

Conclusion

The distinctiveness of nanoscience and nanotechnology, particularly with regard to its popularity in practically all areas of human life, elucidates why the degree of the probable impacts will surpass that of all other conservative and historical technologies. The meeting of the newly rising technologies of the current century has the capability to transform social and economic development and might propose ground breaking and practical solutions for the very critical problems of the global community and its environment. On the other hand, an improved understanding of the prospective benefits and risks of science and technology on a nano-scale is necessary because it will present policy-makers with enhanced tools to make dependable choices. Nanoscience and its developing technologies have the prospects to progress the condition of the developing world, if the applications are engineered and customized to best fit the requirements of the people.

(Anil Varma, Consulting Editor, The Icfai Research Centre Pune. He can be reached at avarma@iupindia.org.)

Index

M

N

O

P

Q

R

S

T

U

V

W

Z